RESIDENTIAL CONSTRUCTION ACADEMY WORKBOOK

CARPENTRY

FIFTH EDITION

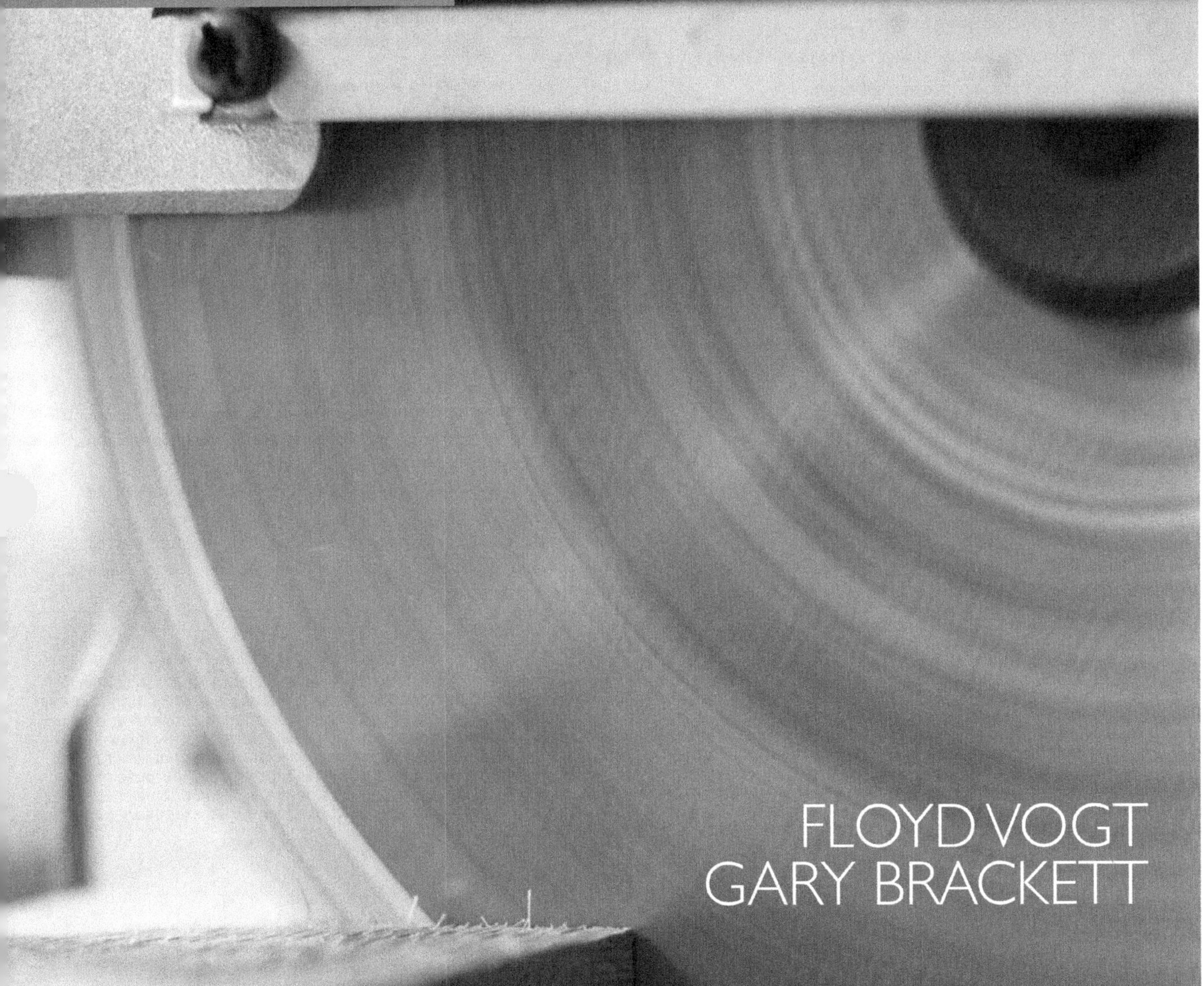

FLOYD VOGT
GARY BRACKETT

CENGAGE

Australia • Brazil • Mexico • Singapore • United Kingdom • United States

CENGAGE

Workbook to Accompany Residential Construction Academy: Carpentry, Fifth Edition
Floyd Vogt and Gary Brackett

SVP, Higher Education & Skills Product: Erin Joyner

Product Director: Matthew Seeley

Senior Product Manager: Vanessa Myers

Product Assistant: Kimberly Klotz

Director, Learning Design: Rebecca von Gillern

Senior Manager, Learning Design: Leigh Hefferon

Learning Designer: Elizabeth Berry

Marketing Director: Sean Chamberland

Marketing Manager: Scott Chrysler

Director, Content Creation: Juliet Steiner

Content Creation Manager: Alexis Ferraro

Content Manager: Jenn Alverson

Digital Delivery Lead: Elizabeth Cranston

Art Director: Jack Pendleton

Text Designer: Erin Griffin

Cover Designer: Erin Griffin

Cover image(s): iStockPhoto.com /stevecoleimages

Library of Congress Control Number: PCN available upon next printing

ISBN: 978-1-3379-1854-1

Cengage
20 Channel Center Street
Boston, MA 02210
USA

Notice to the Reader
Publisher does not warrant or guarantee any of the products described herein or perform any independent analysis in connection with any of the product information contained herein. Publisher does not assume, and expressly disclaims, any obligation to obtain and include information other than that provided to it by the manufacturer. The reader is expressly warned to consider and adopt all safety precautions that might be indicated by the activities described herein and to avoid all potential hazards. By following the instructions contained herein, the reader willingly assumes all risks in connection with such instructions. The publisher makes no representations or warranties of any kind, including but not limited to, the warranties of fitness for particular purpose or merchantability, nor are any such representations implied with respect to the material set forth herein, and the publisher takes no responsibility with respect to such material. The publisher shall not be liable for any special, consequential, or exemplary damages resulting, in whole or part, from the readers' use of, or reliance upon, this material.

Printed at CLDPC, USA, 07-19

Table of Contents

Preface

Designed to accompany *Residential Construction Academy: Carpentry,* Fifth Edition, this workbook provides additional review questions, problems and activities to challenge and reinforce students' comprehension of the content presented in the textbook.

The workbook is divided into chapters, with each chapter directly corresponding to a chapter in the fifth edition of *Residential Construction Academy: Carpentry.* The chapters consist of the features described below, and serve as practice for essential on-the-job skills.

Features

- ***Matching*** questions prompt students to match key terms within the chapter with their definitions to improve their vocabulary for communicating on the jobsite.
- ***True/False*** questions encourage students to consider each description provided and learn to differentiate between accurate and inaccurate information.
- ***Multiple Choice*** questions require students to choose the best answer to complete an accurate statement, thus practicing their knowledge of key concepts in the chapter.
- ***Terms*** provide a list of definitions and require students to fill in the proper term associated with the definition, allowing students another avenue to practice their vocabulary.
- ***Identify*** questions consist of photos and illustrations from the textbook and require students to label each image with the correct name so that students learn how to recognize various materials, tools, equipment and even activities on the jobsite.
- ***List*** activities prompt students to create a list of the specific materials, tools and equipment requested in the question.
- ***Put in Order*** help students recognize the correct sequence for completing a specific hands-on activity performed on the jobsite by requiring them to place each step in the proper order.
- ***Procedures*** describe the specific step-by-step activities presented in the corresponding chapter in the textbook and require students to identify the procedure being described by the correct name.
- ***Activities*** encourage students to apply what they have learned in the corresponding chapter(s) in the textbook and ask them to demonstrate a specific skill or task, thus honing the hands-on skills that are required on the jobsite.
- ***Calculations*** present students with a math problem related to the concepts presented in the corresponding chapter in the textbook, thus enabling them to practice this essential on-the-job skill.

Acknowledgements

Cengage wishes to recognize the individuals who participated in the development of the workbook for the fifth edition of *Residential Construction Academy: Carpentry.*

Special Thanks to Cindy Pickering who developed the questions, problems and activities found within these pages—you have remarkable talent for simplifying even the most complex concepts.

And to James Gallow who performed a technical review of each page in the workbook.

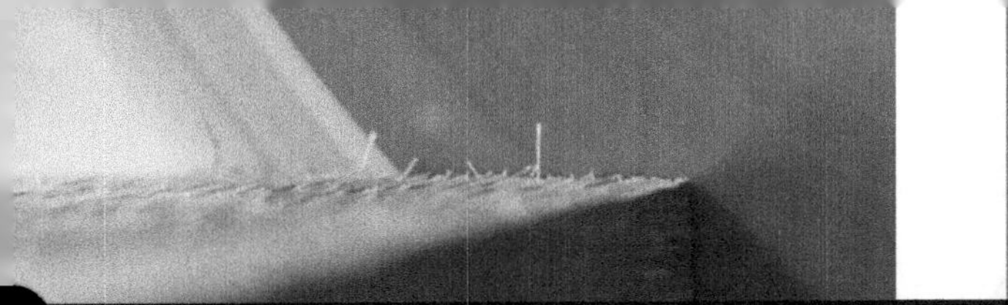

CHAPTER 1

Hand Tools

Matching

Match terms to their definitions. Write the corresponding letters on the blanks. Not all terms will be used.

_______ 1. the back end of objects

_______ 2. an L-shaped cutout along the edge or end of lumber

_______ 3. the honing of a tool by rubbing the tool on a flat sharpening stone

_______ 4. a wide cut, partway and across the grain of lumber

_______ 5. the width of a cut made with a saw

_______ 6. the forward end of tools

_______ 7. vertical: aligned with the force of gravity

_______ 8. a tool used to mark layout and angles, particularly 90-degree angles

_______ 9. a wide cut, partway through and running the grain of lumber

a. crosscut
b. dado
c. groove
d. heel
e. kerf
f. level
g. plumb
h. rabbet
i. square
j. toe
k. whet

True/False

Write *True* or *False* before the statement.

__________ 1. Safety is a blend of ability, skill, and knowledge that should always be applied when working with tools.

__________ 2. A speed square is often called the steel square.

__________ 3. Dropping a carpenter's level could disturb the accuracy of the level.

__________ 4. When unwinding a chalk line, drag it over the surface as you extend the line.

__________ 5. A wet chalk line is best.

__________ 6. The longest block plane is called a jointer.

__________ 7. With aviation snips, the color of the handle denotes the design and use.

__________ 8. A crosscut saw has teeth shaped like rows of tiny chisels.

_______ 9. As a general rule, use nails that are no longer than the thickness of the material being fastened.

_______ 10. If possible, select screws so that two-thirds of their length penetrates the piece in which they are gripping.

Multiple Choice

Choose the best answer. Write the corresponding letter on the blank.

_______ 1. Most carpenters use pocket tapes, which are available in 6-to _____ -foot lengths.

a. 12 c. 35
b. 16 d. 50

_______ 2. The _____ square consists of a movable blade, 1 inch wide and 12 inches long, that slides along the body of the square.

a. combination c. differential
b. framing d. sliding

_______ 3. The shorter of the two legs on a framing square is called the _____.

a. blade c. body
b. tongue d. rafter

_______ 4. The _____ is suspended from a line and hangs absolutely vertical when it stops swinging.

a. plumb bob c. plumb drop
b. level bulb d. level corm

_______ 5. The _____ is used to cut recesses in wood for things such as door hinges and locksets.

a. notch carver c. wood chisel
b. recess carver d. steel chisel

_______ 6. The shortest bench plane is called the _____ plane.

a. jointer c. jack
b. fore d. smooth

_______ 7. _____ handles on aviation snips indicate that they are for straight cuts.

a. Green c. Red
b. Yellow d. Blue

_______ 8. The _____ saw has a narrow blade for making curved cuts of a smaller diameter than the compass saw.

a. coping c. wallboard
b. hack d. keyhole

_______ 9. _____ is the term for nails driven straight.

a. Face nailing
b. Toenailing
c. Top driving
d. Front driving

_______ 10. To extract nails that have been driven home, a _____ is used.

a. wrecking bar
b. flat bar
c. nail claw
d. cat's tail

Terms

Read the description of the uses for particular tools. Write the name of the correct tools on the corresponding blanks.

_______________ 1. Used as a compass to lay out circles and arcs and as dividers to space off equal distances

_______________ 2. Used to cut thin metal, such as roof flashing and metal roof edging

_______________ 3. Universal cutting tool often used for cutting gypsum board, softboards, and a variety of finish materials

_______________ 4. Designed especially for gypsum board

_______________ 5. Used to set nailheads below the surface for finishing

_______________ 6. A bar used to pry small work and pull small nails

_______________ 7. Open like scissors and are a quick way to hold material together

_______________ 8. Hardwood blocks with threaded rods through them that are used for holding on a large surface area without marring

_______________ 9. Useful for holding objects together while they are being fastened, holding temporary guides, and applying pressure to glued joints

_______________ 10. Clamps designed for speed and ease of operation

Identify

Identify the following tools used for carpentry. Write the correct name on the line next to the item.

1. ______________________________

2. ______________________________

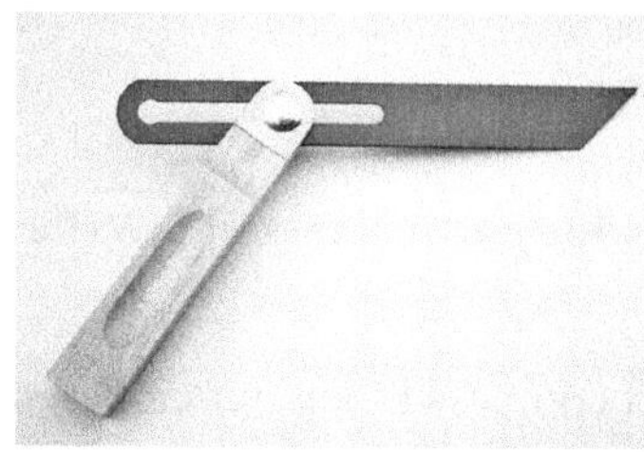

3. ______________________________

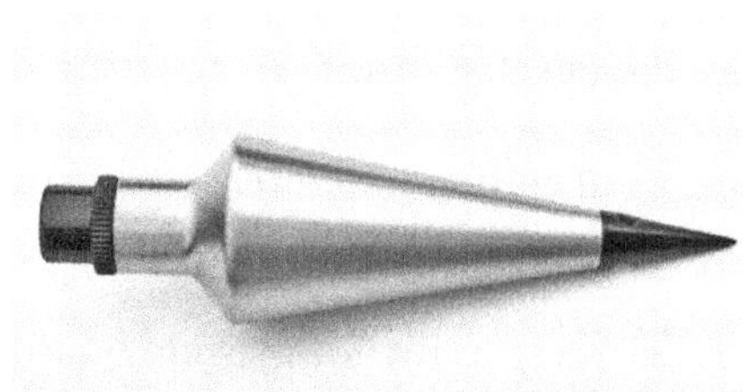

4. ______________________________

5. ______________________________

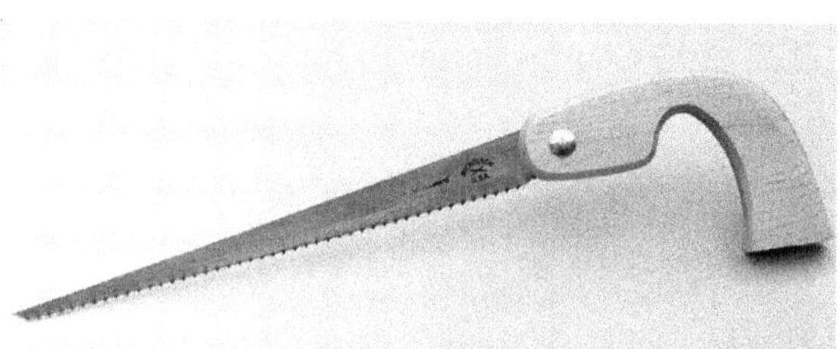

6. ______________________________

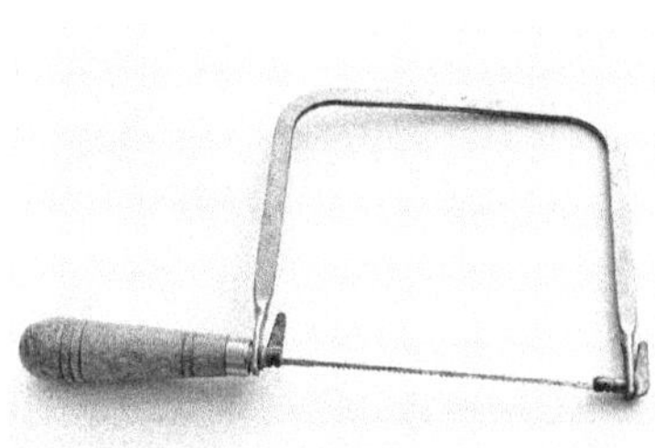

7. ______________________________

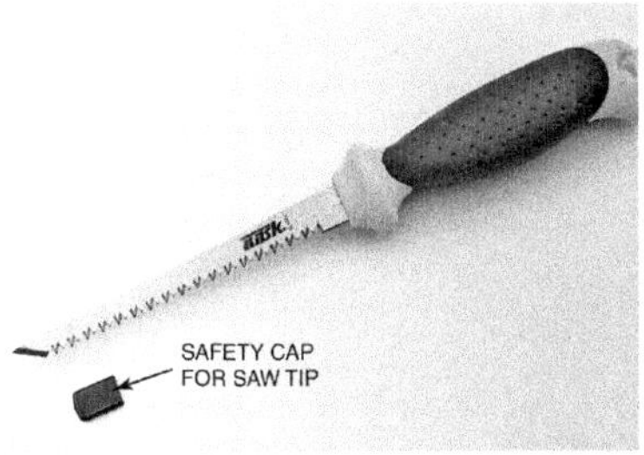

8. ______________________________

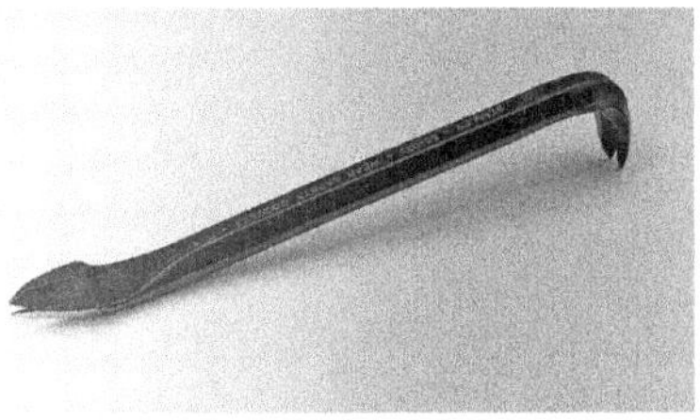

9. ______________________________

List

List four tooth-cutting tools on the lines below.

1. ______________________________
2. ______________________________
3. ______________________________
4. ______________________________

CHAPTER 2

Hand Power Tools

Matching

Match terms to their definitions. Write the corresponding letters on the blanks. Not all terms will be used.

______ 1. guide for ripping lumber on a table saw

______ 2. sloping edge or side of a piece at any angle other than a right angle

______ 3. cutting of the end of a piece at any angle other than a right angle

______ 4. edge or end bevel that does not go all the way across the edge or end

______ 5. bevel cut across the width and also through the thickness of a piece

a. bevel
b. chamfer
c. compound miter
d. fence
e. miter
f. gate
g. plunge cut

True/False

Write *True* or *False* before the statement.

__________ 1. Safe operation of power tools requires knowledge and discipline.

__________ 2. On a circular saw, loosen the bolt that holds the blade in place by rotating the bolt in the opposite direction as the rotation of the blade.

__________ 3. Blades with more teeth cut faster but rougher.

__________ 4. Some reciprocating saws can be switched to several orbital cutting strokes from a straight back-and-forth cutting action.

__________ 5. Fast speeds are desirable when using a power drill to make large holes or when drilling metal.

Multiple Choice

Choose the best answer. Write the corresponding letter on the blank.

______ 1. A ____ saw is used primarily for roughing in work.

a. saber
b. reciprocating
c. circular
d. Skilsaw

______ 2. Common blade lengths for a reciprocating saw run from ____ inches.

a. 4 to 12
b. 5 to 10
c. 3 to 15
d. 4 to 14

_______ 3. _____ drills usually have a pistol grip handle.

a. Carpentry-grade c. Heavy-duty
b. Professional-grade d. Light-duty

_______ 4. The size of a drill is determined by the _____.

a. capacity of the chuck c. required trigger switch pressure
b. maximum rpm possible d. design of the body and handle

_______ 5. When boring, power drill speeds should be _____ when drilling wood with moderate pressure on the material.

a. very low c. fairly high
b. moderate d. extremely high

_______ 6. _____ are extensively used for fastening gypsum board to walls and ceilings.

a. Pneumatic impact tools c. Hammer-drills
b. Impact drivers d. Screw guns

_______ 7. An adjustable _____ on a power plane allows planing square, beveled edges to 45 degrees.

a. gate c. edge
b. fence d. hedge

_______ 8. A light-duty specialized type of router called a _____ is used almost exclusively for trimming back the edge of plastic laminates.

a. laminate trimmer c. facade trimmer
b. surface edger d. coating edger

_______ 9. A(n) _____ sander is frequently used for sanding cabinetwork and interior finish.

a. orbital finishing c. belt
b. oscillating finishing d. random orbital

_______ 10. Powder charges for powder-actuated drivers are _____ according to strength.

a. color-coded c. different sizes
b. numerically-stamped d. alphabetically-marked

Put in Order

Put in order the steps for making plunge cuts with a circular saw. Write the corresponding numbers on the blanks.

_______________ A. Hold the guard open and tilt the saw up with the front edge of the base resting on the work.

_______________ B. Making sure the teeth of the blade are not touching the work, start the saw.

_______________ C. Lay out the cut to be made. Wear eye and ear protection. Adjust the saw for the depth of the cut.

_______________ D. Move the saw blade over, and in line with, the cut to be made.

_______________ E. Lower the blade slowly into the work by rotating the saw with the front edge of the base as a pivot.

Identify

Identify the following power tools and accessories. Write the correct name on the line with the corresponding number next to the images.

Courtesy of Porter Cable.

1. ______________________________

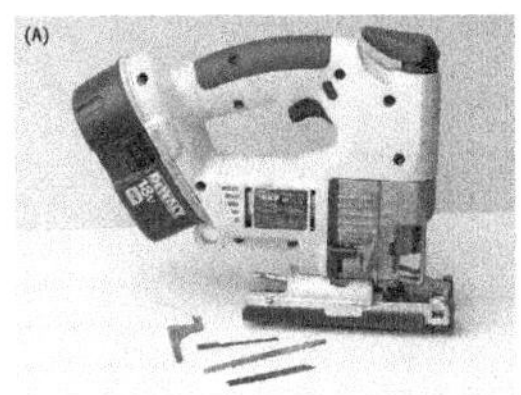

2. ______________________________

3. ______________________________

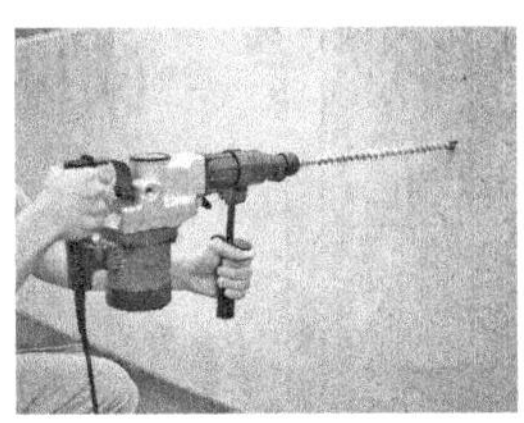

4. ______________________________

5. ______________________________

Courtesy of Robert Bouch Tool Corporation.

6. ______________________________

7. ______________________________

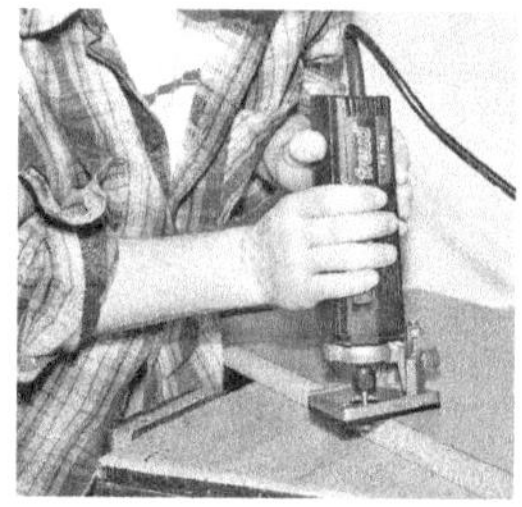

8. ______________________________

Courtesy of DEWALT

9. ______________________________

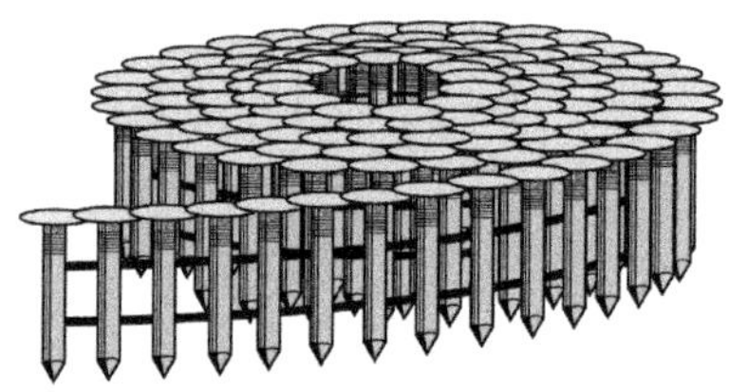

Courtesy of Palosade.

10. ______________________________

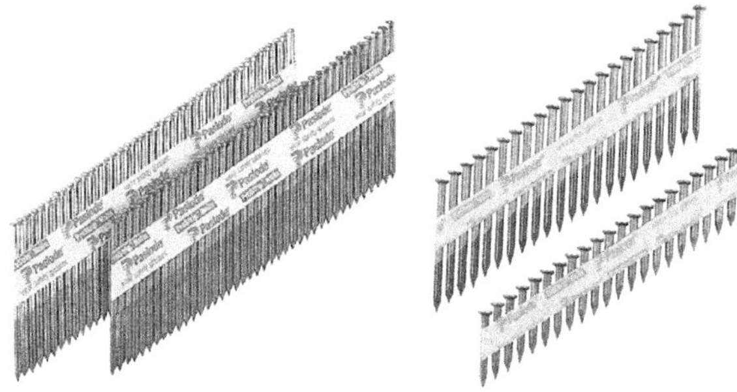

Courtesy of Palosade.

11. ______________________________

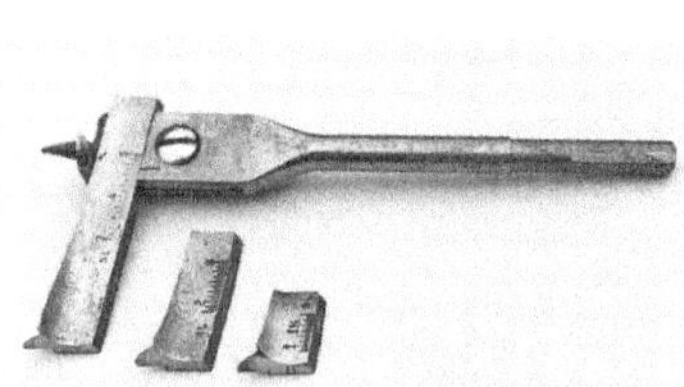

12. ______________________________

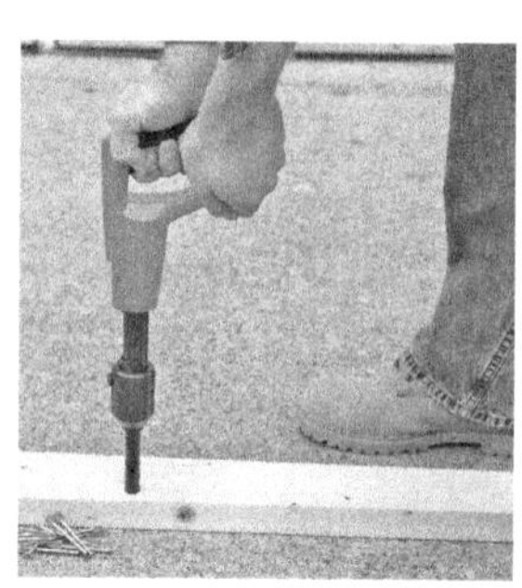

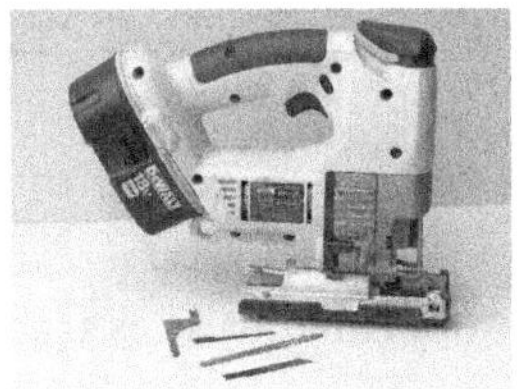

13. ______________________________

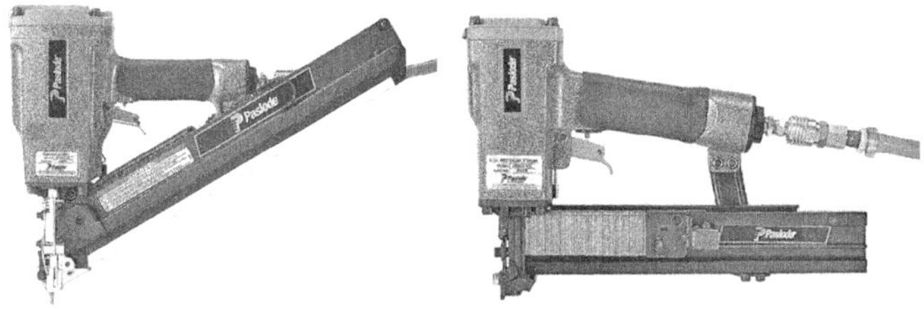

Courtesy of Palosade.

14. ______________________________

List

List five of the general safety rules for operating hand power tools. Write the correct answers on the corresponding lines.

1. ______________________________
2. ______________________________
3. ______________________________
4. ______________________________
5. ______________________________

Procedures

Read the following procedures. Determine what procedure is being described and write the name of the procedure on the blank.

Procedure 1: ______________________________

- Outline the cut to be made. Wear eye and ear protection.
- Secure the piece being planed.
- Set the side guide to the desired angle, and adjust the depth of cut.
- Hold the toe (front) firmly on the work, with the plane cutterhead clear of the work.
- Start the motor. With steady, even pressure, make the cut through the work for the entire length.
- Guide the angle of the cut by holding the guide against the side of the stock.
- Apply pressure to the toe of the plane at the beginning of the cut. Apply even pressure after the heel is on the piece. Move pressure to the heel (back) at the end of the cut to prevent tipping the plane over the ends of the work.

Procedure 2: ____

- Wear eye, respiratory, and ear protection.
- Secure the work to be sanded.
- Make sure the belt is centered on the rollers and is tracking properly.
- Holding the tool with both hands, start the machine.
- Place the pad of the sander flat on the work.
- Pull the sander back and lift it just clear of the work at the end of the stroke.
- Bring the sander forward. Continue sanding using a skimming motion that lifts the sander just clear of the work at the end of every stroke. Sanding in this manner prevents overheating the sander, the belt, and the material being sanded. It allows debris to be cleared from the work. The operator can also see what has been done.
- Do not sand in one spot too long. Be careful to keep the sander flat, not tilting the sander in any direction. Always sand with the pad flat on the work. Do not exert excessive pressure. The weight of the sander is enough. Always sand with the grain to produce a smooth finish.
- Make sure the sander has stopped before setting it down. It is a good idea to lay it on its side to prevent accidental traveling.

Procedure 3: ____

- Study the manufacturer's directions for safe and proper use of the gun.
- Wear eye and ear protection.
- Make sure the drivepin will not penetrate completely through the material into which it is driven. This has been the cause of fatal accidents.
- To prevent ricochet hazard, make sure the recommended shield is in place on the nose of the gun. Many different shields are available for special fastening jobs.
- Select the proper fastener for the job. Consult the manufacturer's drivepin selection chart to determine the correct fastener size and style.
- Select a powder charge of necessary strength. Always start with the weakest charge that will do the job. Load the driver with the pin first and the cartridge second. Keep the tool pointed at the work.
- Press the tool hard against the work surface, and pull the trigger. The resulting explosion drives the pin.
- Eject the spent cartridge and clean the tool as needed.

Procedure 4: ____

- Wear eye, respiratory, and ear protection.
- Select the desired grit sandpaper. Attach it to the pad, making sure it is tight. A loose sheet will tear easily.
- Start the motor and sand the surface evenly, slowly pushing and pulling the sander with the grain. Let the action of the sander do the work. Do not use excessive pressure as this may overload the machine and burn out the motor.
- Always hold the sander flat on its pad.

Procedure 5: ____

- Wear eye and ear protection.
- Select the correct bit for the type of cut to be made.

- Insert the bit into the chuck. Make sure the chuck grabs at least twice the shaft diameter of the bit. Adjust the depth of cut.
- Clamp the work securely in position. Plug in the cord.
- Lay the base of the router on the work with the router bit clear of the work. Start the motor.
- Advance the bit into the cut, pulling the router in a direction that is against the rotation of the bit. To rout the outside edges and ends, the router is moved counterclockwise around the piece. When making internal cuts, the router is moved in a clockwise direction.

Activities

Under the supervision of your instructor and observing all safety regulations and rules perform the following power tool operations.

When you can perform each of these operations correctly and understand all rules about each tool, initial the form. Your instructor will initial each operation when he/she is satisfied with your competence. When he/she is satisfied that you are safe and knowledgeable on that tool, you will be allowed to use that tool to work in the H.B.I. shop or jobsite.

	Date	Instructor Initials	Student Initials
CIRCULAR SAW			
Install saw blade and adjust for depth and angle			
Crosscuts			
Crosscutting with a bevel			
Steep angle cuts			
Steep angle cuts with bevel			
Cut sheet goods			
Plunge cuts			
Cut with rip guide			
SABER (JIG) SAW			
Install a saw blade			
Adjust the base for an angle			
Make a cut and cut shape			
Make a plunge cut			
RECIPROCATING SAW (SAWZALL)			
Install blade			
Make a cut			
ELECTRIC DRILL			
Install a twist drill bit and drill a hole			
Install a spade bit and drill a hole			
Install an auger bit and drill a hole			
Install a hole saw and drill a hole			
SCREW GUN			
Remove and reinstall PT2 tip			
Adjust head to proper depth for drywall			
Install 100 1/4" and 10 1/2" screws flush			
POWER PLANER			
Review parts of planer			
Make thickness setting			
Plane a surface			

	Date	Instructor Initials	Student Initials
ROUTER			
Install bit and collet and adjust depth			
Set up and rout a dado			
Set up and rout a rabbet			
Set up and rout with an edge-guided bit			
Use edge guide			
BELT SANDER			
Remove and install a sanding belt			
Adjust belt tension			
Sand a surface			
Clean belt cleaner			
FINISH SANDER			
Change paper			
Sand block wood to finish readiness			
PNEUMATIC NAILERS			
Load fasteners in magazine			
Make a trial fastening			
Adjust air pressure as needed			
Single fire and rapid fire fastening			
POWDER-ACTUATED DRIVERS			
Review safety protocol of device			
Select and load fastener			
Select and load charge			
Position material and driver for fastening			
Pull trigger and reload for next fastening			
EXTENSION POWER CORDS AND GFCI			
Inspect power cord			
Correctly connect power tool to extension power cord and GFCI			
Inspect GFCI			

CHAPTER 3

Stationary Power Tools

Matching

Match terms to their definitions. Write the corresponding letters on the blanks. Not all terms will be used.

_______ 1. a guide for ripping lumber on a table saw

_______ 2. sawing lumber in the direction of the grain

_______ 3. a cut made across the grain of lumber

_______ 4. a guide used on the table saw for making miters and square ends

a. crosscut

b. fence

c. miter gauge

d. rip

e. shaper

True/False

Write *True* or *False* before the statement.

____________ 1. The band saw efficiently cuts curves in a vertical motion similar to the jig saw.

____________ 2. When using a table saw, stand directly in back of the saw blade.

____________ 3. Leaving the cut stock between the fence and a running saw blade may cause a kickback.

____________ 4. The miter gauge should be used at the same time with the rip fence.

____________ 5. Backing out of long cuts with a band saw could cause the band to come off the drive wheels.

____________ 6. When using a jointer, check the manufacturer's recommendations for the maximum allowable depth to cut.

____________ 7. Use your fingers to hold the stock directly over the cutters when using a jointer.

____________ 8. If the bit binds in the hole and the work has not been secured, the entire piece may spin out of control.

____________ 9. The fence on a shaper provides for a quick setup for straight edges such as a trim.

____________ 10. Bench sanders are used to sand materials slowly and meticulously.

Multiple Choice

Choose the best answer. Write the corresponding letter on the blank.

_______ 1. A table saw may measure up to _____ inches.

a. 10 c. 14
b. 12 d. 16

_______ 2. Blades are secured to a table saw with a(n) _____ nut.

a. arbor c. closed
b. binding d. stopper

_______ 3. A _____ set is used to cut only partway through stock thickness during a crosscutting operation.

a. guard c. dado
b. stock d. miter

_______ 4. The miter saw is also called a _____.

a. compound box saw c. power miter box
b. crown molding saw d. sliding compound box

_______ 5. Which of the following statements about a band saw is correct?

a. The band saw is not designed to cut irregular shapes.
b. The band saw is usually operated freehand.
c. The band saw has one large wheel.
d. The band saw size is determined by the width of the band.

_______ 6. _____ are performed on the jointer by adjusting the fence over the cutterhead such that only a small portion of it is available for cutting.

a. Chamfers c. Collars
b. Compound miters d. Rabbets

_______ 7. A _____ is a heavy-duty router attached upside down to a base.

a. shaper c. compound miter saw
b. jointer d. sliding miter saw

_______ 8. The spindle shaft of a shaper is typically _____ diameter and spins about 9,000–18,000 RPM.

a. ¼ inch c. ¾ inch
b. ½ inch d. ⅛ inch

_______ 9. There are _____ different styles of bits available for the shaper.

a. twelve c. thirty
b. twenty-seven d. hundreds of

_______ 10. _____ are designed to smooth almost any shaped piece of work.

a. Bench sanders c. Table saws
b. Jointers d. Sliding compound miter saws

Put in Order

Put in order the steps for crosscutting stock to length with a power miter saw. Write the corresponding numbers on the blanks.

_______________ A. Place and hold firmly the stock against the fence with one hand, which should be out of the line of cut by as much as possible.

_______________ B. Start the motor and push the saw gently down into the material cutting the waste side of the layout line.

_______________ C. Unlock the storage clamps and allow the saw to rise up into ready position. Lay out the desired length. Wear eye and ear protection.

_______________ D. Release the trigger and allow the saw to come to the rest position again.

_______________ E. If using a sliding model, push the saw gently back to finish the cut.

_______________ F. If using a sliding model, with the other hand pull the saw forward and down close to the material.

Terms

Read the description of the uses for particular tools. Write the name of the correct tools on the corresponding blanks.

_______________ 1. Used to crosscut and rip by pushing material through the saw

_______________ 2. Cuts in a chopping action at many angles for joining many finish materials

_______________ 3. Used to cut in a continuous downward action allowing curved cuts

_______________ 4. Smooths material edge surfaces by passing material over the cutter

_______________ 5. Designed for making precise holes

_______________ 6. Allows repetitive smoothing of edges and surfaces

Identify

Identify the following stationary power tools. Write the correct name on the line next to the item.

1. ______________________________

2. ______________________________

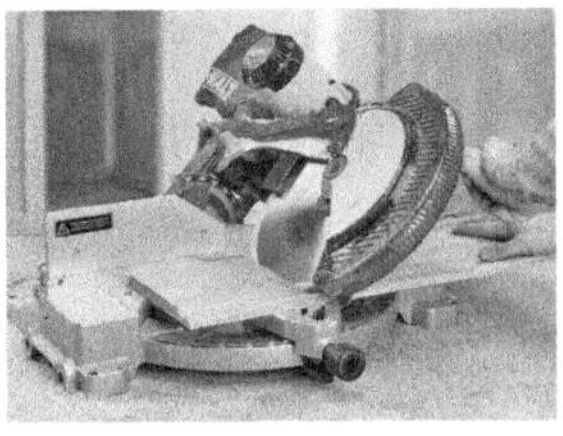

Courtesy of DeWalt.

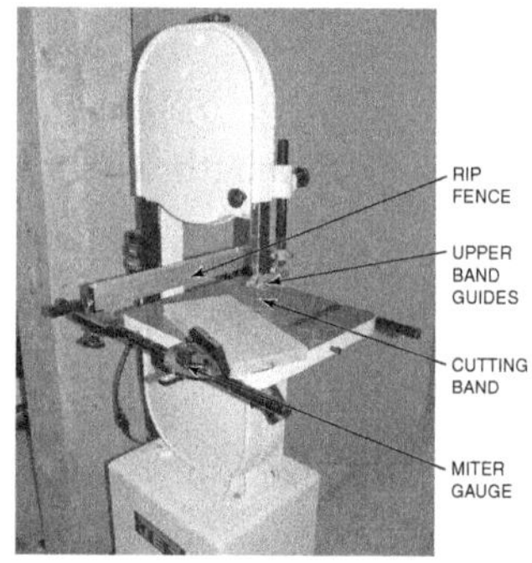

3. ______________________________

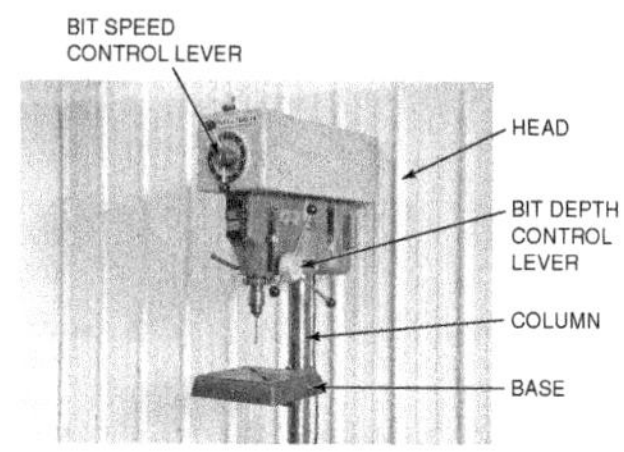

4. ______________________________

Courtesy of Laguna Tools

5. ______________________________

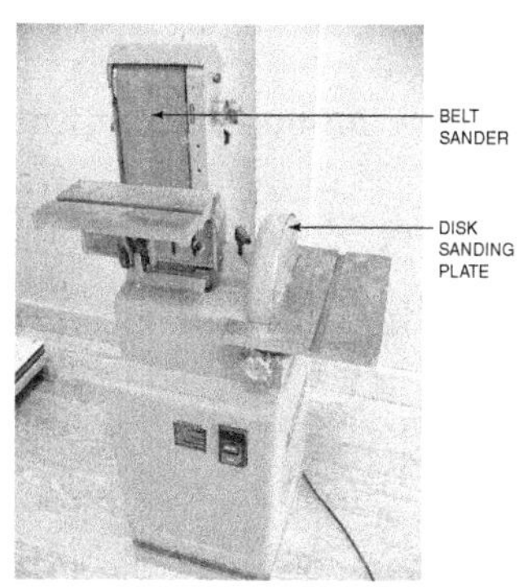

6. ______________________________

List

List five general safety rules for operating stationary power tools. Write the correct answers on the corresponding lines.

1. ______________________________
2. ______________________________
3. ______________________________
4. ______________________________
5. ______________________________

Procedures

Read the following procedures. Determine what procedure is being described and write the name of the procedure on the blank.

Procedure 1: ____

- Wear eye and ear protection.
- Measure from the rip fence to the point of a saw tooth set closest to the fence. Lock the fence in place. Check and adjust the rip fence measuring scale, if necessary.
- Adjust the height of the blade to about ¼ inch above the stock to be cut. Some manufacturers recommend setting the blade at full height inside the blade guard to allow the blade to run cooler and cut more easily.
- With the stock clear of the blade, turn on the power.
- Hold the stock against the fence with the left hand. Push the stock forward with the right hand, holding the end of the stock.
- Push the stock firmly, listening to the saw to determine appropriate feed speed. The blade speed should always be allowed to run at full.
- As the end approaches the saw blade, let the stock slip through the left hand, removing it from the work. If the stock is of sufficient width (at least 5 inches wide), finish the cut with the right hand pushing the end all the way through the saw. Otherwise use a *push stick* to finish the cut.

Procedure 2: ____

- Lay out the desired length of stock.
- Wear eye and ear protection.
- Set desired blade and miter gauge angles.
- Hold stock firmly against miter gauge back from the blade.
- Turn on motor and, using two hands, ease the stock into the blade.
- Stand to the side of the line of the blade.
- Finish cut and remove the stock before returning the miter gauge to the start position.
- Keep the blade clear of debris.

Procedure 3: ____

- Safe operation requires that the operator secure the stock with two hands. Begin with the stock resting on the infeed table and fence.
- Concentrate on keeping the stock against the table and fence during the entire pass, preventing it from wobbling.
- Then slowly ease the stock toward the safety shield and into the cutter head.
- At some point halfway through the cutting pass, the operator's hands must shift to secure the stock to the outfeed table and fence.
- Take care to keep the stock moving and firmly held against the fence.

Activities

Under the supervision of your instructor and observing all safety regulations and rules perform the following power tool operations.

When you can perform each of these operations correctly and understand all rules about each tool, initial the form. Your instructor will initial each operation when he/she is satisfied with your competence. When he/she is satisfied that you are safe and knowledgeable on that tool, you will be allowed to use that tool to work in the H.B.I. shop or jobsite.

	Date	Instructor Initials	Student Initials
TABLE SAW			
Adjust fence	________	________	________
Lockout and change saw blade	________	________	________
Install splitter	________	________	________
Rip and bevel a board	________	________	________
Crosscut a board	________	________	________
Set up and cut a dado and rabbet	________	________	________
MITER SAW			
Make a straight cut	________	________	________
Cut miter	________	________	________
Cut a miter to long point	________	________	________
Cut a miter to short point	________	________	________
Cut a bevel	________	________	________
Cut a compound miter	________	________	________
BAND SAW			
Adjust upper guide to proper height	________	________	________
Cut stock using a miter guide	________	________	________
Cut stock using a rip fence	________	________	________
Cut stock free hand following a curved line	________	________	________
JOINTER			
Adjust fence and infeed table	________	________	________
Joint an edge	________	________	________
Joint a taper	________	________	________
DRILL PRESS			
Adjust table level	________	________	________
Set up chuck and check operation	________	________	________
Drill a hole	________	________	________
SHAPER TABLE			
Assemble shaper	________	________	________
Install bit	________	________	________
Cut rabbet	________	________	________
Cut dado	________	________	________
Make 1/4 round	________	________	________
BENCH SANDER			
Square disc platform	________	________	________
Change disc	________	________	________
Adjust belt	________	________	________
Sand a piece of wood	________	________	________

CHAPTER 4

Wood and Wood Products

Matching

Match terms to their definitions. Write the corresponding letters on the blanks. Not all terms will be used.

_______ 1. a layer just inside the bark of a tree where new cells are formed

_______ 2. a process in which shorter lengths are glued together using deep, thin V grooves, resulting in longer lengths

_______ 3. trees that shed leaves each year

_______ 4. technique of removing water from lumber using natural wind currents

_______ 5. framing members placed at right angles to joists, studs, and rafters to form and support openings

_______ 6. the small, soft core at the center of a tree

_______ 7. the outer part of a tree just beneath the bark containing active cells

_______ 8. trees that are cone-bearing; also known as evergreen trees

_______ 9. a measure of lumber volume

_______ 10. a warp in a board along its length forming an arc

_______ 11. a method of sawing lumber that produces a flat-grain where annular rings tend to be parallel to the width of the board

_______ 12. treated in a special way to make a material harder and stronger

_______ 13. the rings seen when viewing a cross-section of a tree trunk

a. air-drying
b. annular rings
c. board foot
d. cambium layer
e. coniferous
f. crown
g. deciduous
h. dimension
i. dry kiln
j. finger joint
k. header
l. heartwood
m. medullary rays
n. millwork
o. panel
p. pith
q. plain-sawed
r. quarter-sawed
s. sapwood
t. tempered

True/False

Write *True* or *False* before the statement.

_____________ 1. Engineered lumber has reduced manufacturing waste but products have less strength than traditional solid lumber products.

_____________ 2. Manufacturing wood products from logs uses less energy than manufacturing metal or masonry products.

_______________ 3. Lumber used in framing should not have any warping.

_______________ 4. Plywood manufacturing process uses the best quality veneer logs for inner cores.

_______________ 5. Engineered lumber can be slippery compared to standard lumber.

Multiple Choice

Choose the best answer. Write the corresponding letter on the blank.

_______ 1. The long, narrow surface of a piece of lumber is called its _____ .

a. side c. edge
b. width d. length

_______ 2. A(n) _____ is a mild type of warp that is an arc along the length of the board.

a. wave c. curl
b. arch d. crown

_______ 3. _____ is the classification for the strongest softwood veneers.

a. Group 1 c. Grade A
b. Group 3 d. Grade C

_______ 4. _____ performance-rated panels are intended for use when long delays in construction may cause the panels to be exposed to the weather unprotected.

a. Exposure 1 c. Exterior 2
b. Exposure A d. Exterior B

_______ 5. MDF would most likely be used for _____.

a. decorative panels c. ceiling tiles
b. cabinet doors d. interior wall paneling

Terms

Read the descriptions of various wood products. Write the name of the correct product on the corresponding blanks.

_______________ 1. Reconstituted wood panel made of wood flakes, chips, sawdust, and planer shavings

_______________ 2. Manufactured in a manner similar to that used to make hardboard except that the fibers are not pressed as tightly together

_______________ 3. Also called softboard; used primarily for insulating or sound control purposes

_______________ 4. Engineered lumber intended for use as high-strength, load-carrying beams to support the weight of construction over window and door openings

_______________ 5. Designed to replace large dimension lumber (beams, planks, and posts)

_______________ 6. Engineered lumber used for a wide range of millwork that requires high-grade lumber

_______________ 7. Nonveneered performance-rated structural panel with the same strength and stability as plywood

_______________ 8. Assemblies that carry heavy loads over long distances while using considerably less wood than solid lumber

_______________ 9. Constructed of solid lumber glued together, side against side; used for structural purposes yet are decorative as well

Identify

Identify the following images related to wood and wood products. Write the correct name on the line next to the item.

1. ______________________________________

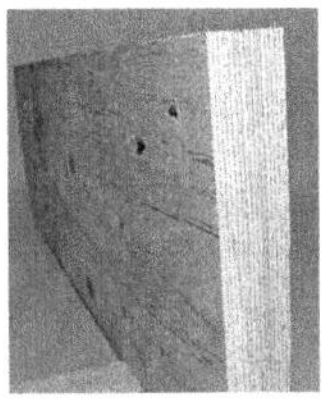

2. ______________________________________

3. ______________________________________

4. ______________________________________

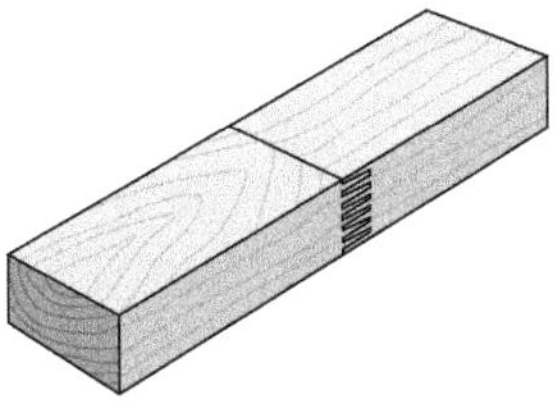

5. ______________________________________

Courtesy of Louisiana Pacific Corporation.

List

List three kinds of wood warp. Write answers on the lines below.

1. ______________________________

2. ______________________________

3. ______________________________

Calculations

Use the information provided in the table to calculate board feet. Write answers in the last column.

	Number of Pieces	Length (ft)	Width (in)	Thickness (in)	Number of Board Feet
1.	23	8	6	1	
2.	12	10	8	2	
3.	45	8	4	2	
4.	5	12	10	1	
5.	120	8	2	2	
6.	37	10	10	2	
7.	56	12	12	1	
8.	12	8	6	2	
9.	25	8	4	1	

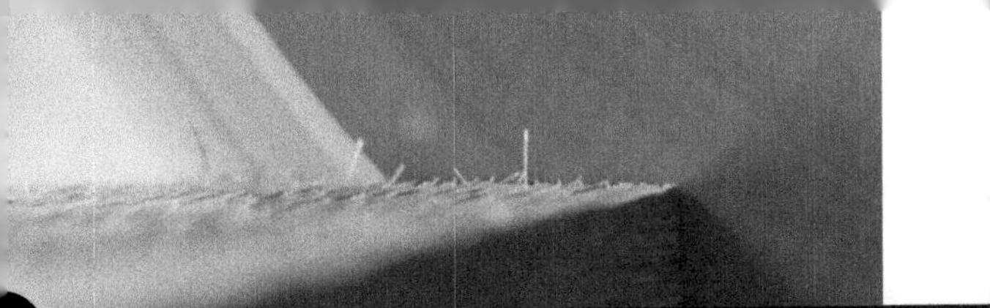

CHAPTER 5

Fasteners

Matching

Match terms to their definitions. Write the corresponding letters on the blanks. Not all terms will be used.

_______ 1. a device used to fasten structural member in place

_______ 2. a thick adhesive

_______ 3. a double-headed nail used for temporary fastening

_______ 4. method of driving a nail straight through a surface material into a supporting member

_______ 5. protected from rusting by a coating of zinc

_______ 6. a thin, short, finishing nail

_______ 7. a term used in designating nail sizes

_______ 8. the decomposition of one of two unlike metals in contact with each other in the presence of water

_______ 9. method of driving a nail diagonally through a surface material into a supporting member

a. anchor
b. box nail
c. brad
d. duplex nail
e. electrolysis
f. face nail
g. finish nail
h. galvanized
i. mastic
j. penny (d)
k. toenail

True/False

Write *True* or *False* before the statement.

____________ 1. Nails can be hard and brittle at the same time.

____________ 2. In the penny system, the shortest nail is 6d and 1 inch long.

____________ 3. Masonry nails are designed to bend around stones in concrete.

____________ 4. Screw lengths are measured from the point to that part of the head that sets flush with the wood when fastened.

____________ 5. Carriage bolts installed in wood should be tightened as firmly as possible.

____________ 6. The disadvantage of using a toggle bolt is that, if removed, the toggle falls off inside the wall.

____________ 7. Hollow wall expansion anchors are commonly called molly screws.

____________ 8. Connectors are plastic pieces formed into various shapes to join wood to wood.

____________ 9. Glue is mostly used for interior finish work.

____________ 10. Pieces glued with contact cement must be clamped under pressure.

Multiple Choice

Choose the best answer. Write the corresponding letter on the blank.

_______ 1. Uncoated steel nails are called _____ nails.

a. matte
b. galvanized
c. shiny
d. bright

_______ 2. Most nails, cut from long rolls of metal wire, are called _____ nails.

a. rolled
b. wire
c. cut
d. spun

_______ 3. A _____ nail is often used on temporary structures, such as wood scaffolding and concrete forms because it is easy to pry out.

a. duplex
b. casing
c. brad
d. box

_______ 4. The _____ bolt has a square section under its oval head, and the square section is embedded into the wood.

a. machine
b. stove
c. carriage
d. common

_______ 5. The _____ anchor is used when high resistance to pullout is required.

a. wedge
b. sleeve
c. drop-in
d. split-fast

_______ 6. The _____ is used with a lag screw and is expanded as the fastener is threaded in.

a. double expansion anchor
b. single expansion anchor
c. lag shield
d. concrete shield

_______ 7. The deep threads of the _____ anchor resist stripping out when screwed into gypsum board, strand board, and similar material.

a. toga bolts
b. universal plug
c. molly bolt
d. conical screw

_______ 8. _____ is faster setting, so joints should be made quickly after applying the glue.

a. Brown glue
b. Yellow glue
c. Contact cement
d. Rubber cement

_______ 9. _____ is used in a glued floor system and can be used on wet or frozen wood.

a. Construction adhesive
b. Panel adhesive
c. Laminate mastic
d. Wood glue

_______ 10. _____ help fasten tops and bottoms of posts and columns.

a. Caps and bases
b. Framing ties
c. Hangers
d. Anchors

Terms

Read the following descriptions of fasteners. Write on the corresponding blanks the name of the fastener being described.

_______________ 1. Made of heavy gauge wire and have a medium-sized head with a barbed section just under it

_______________ 2. Light-gauged with a very small head, which gets sunk into the wood

_______________ 3. Used extensively to fasten metal framing and has a cutting edge to eliminate predrilling a hole

_______________ 4. Have a square or hex head; designed to be turned with a wrench

_______________ 5. Consists of an expansion shield and a cone-shaped, internal expander plug

_______________ 6. One-piece steel with two sheared expanded halves at the base

_______________ 7. Made of nylon and is used for a number of hollow wall and some solid wall applications

_______________ 8. Used to apply wall paneling, foam insulation, gypsum board, and hardboard to wood, metal, and masonry

Identify

Identify the following fasteners. Write the correct fastener name on the line next to the corresponding number or letter.

1. A. ______________________________

   B. ______________________________

   C. ______________________________

   D. ______________________________

   E. ______________________________

   F. ______________________________

   G. ______________________________

   H. ______________________________

   I. ______________________________

   J. ______________________________

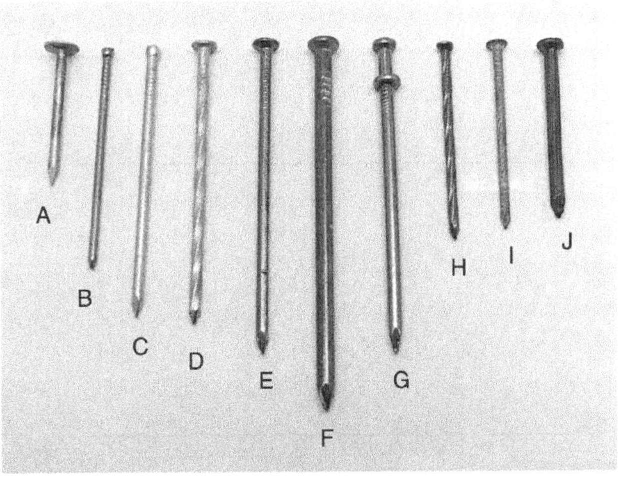

2. A. ______________________________

   B. ______________________________

   C. ______________________________

   D. ______________________________

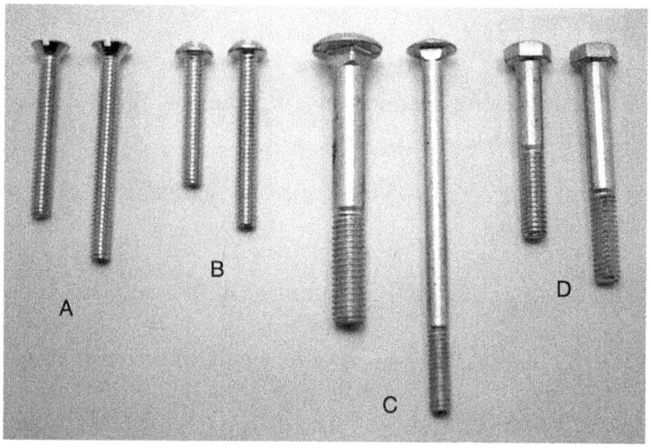

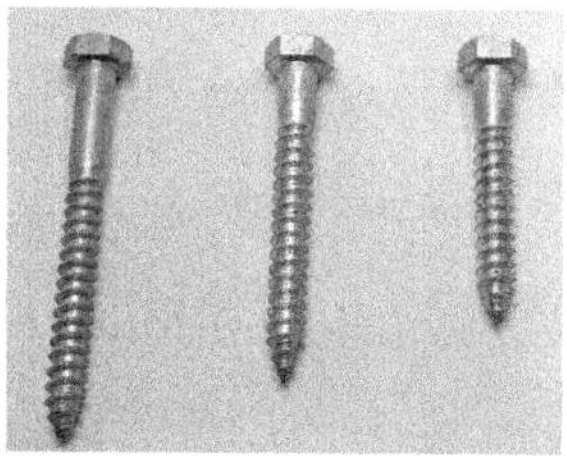

3. ______________________________

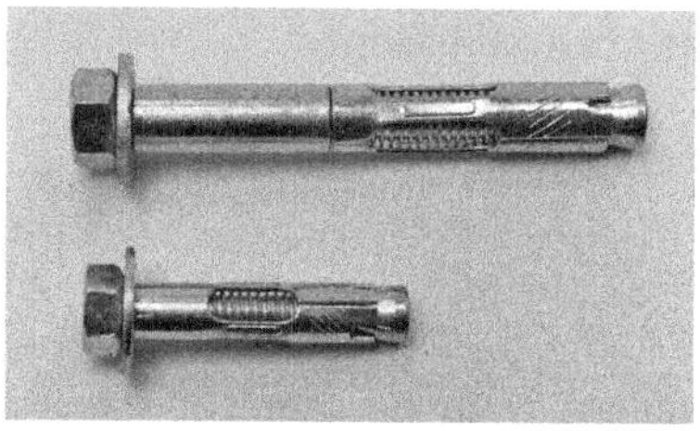

4. ______________________________

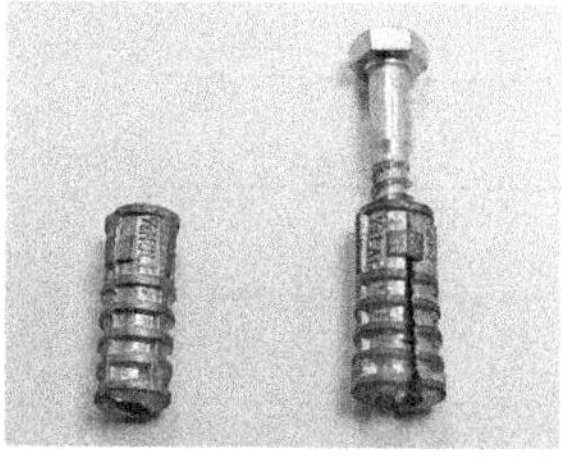

5. ______________________________

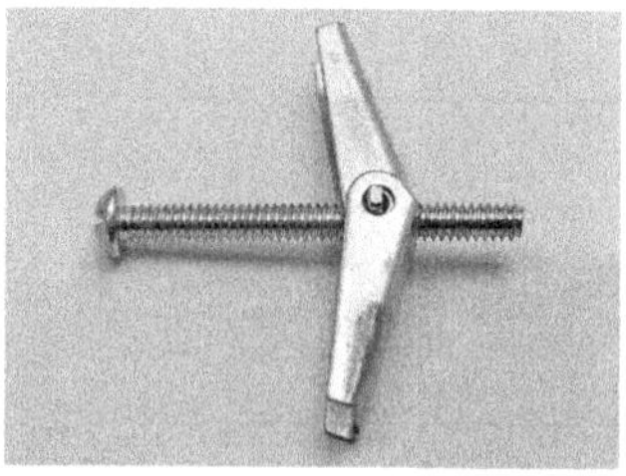

6. ______________________________

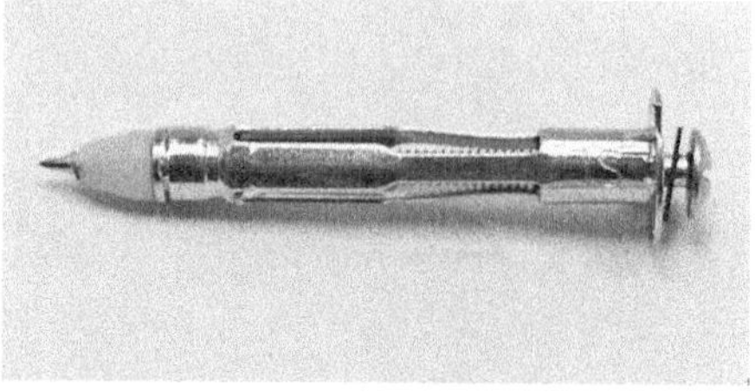

7. ______________________________

List

List three types of wood-to-concrete connectors. Write answers on the lines below.

1. ____

2. ____

3. ____

List three types of wood-to-wood connectors. Write answers on the lines below.

4. ____

5. ____

6. ____

CHAPTER 6

Jobsite Safety and Construction Aids

Matching

Match terms to their definitions. Write the corresponding letters on the blanks. Not all terms will be used.

______ 1. worker whose responsibilities include safe assembly of scaffolding

______ 2. designated person on a jobsite who is capable of identifying hazardous situations and has authority to take prompt corrective measures to eliminate them

______ 3. heavy wood blocks and framing used as a foundation for scaffolding

______ 4. a small strip of wood applied to support a shelf or similar piece

______ 5. person who works on scaffolding

a. cleat

b. competent person

c. crib

d. erector

e. user

f. agent

g. client

True/False

Write *True* or *False* before the statement.

__________ 1. Laser light from laser leveling devices can cause permanent eye damage.

__________ 2. Dry cutting of dust from masonry work produces silica dust that may cause lung diseases years later.

__________ 3. Materials should be unpacked from shipping containers upon arrival at the jobsite.

__________ 4. The bottom legs of the scaffold will not be overloaded without overloading the bearer or top horizontal member of any frame.

__________ 5. The rolling tower, or mobile scaffold, is widely used for small jobs and is generally not more than 20 feet in height

Multiple Choice

Choose the best answer. Write the corresponding letter on the blank.

______ 1. According to OSHA's guidelines, which of the following exposures has the greatest risk for hearing damage?

a. Running a backhoe for six hours

b. Driving a bulldozer for two hours

c. Running a circular saw for 45 minutes

d. Using a chain saw for 15 minutes

_______ 2. Which of the following hardhat classes provides the best protection from electrical hazards?

a. Class A
b. Class B
c. Class C
d. Both A and C

_______ 3. Excavations require adequate cave-in protection if they are _____.

a. more than 3 feet wide
b. less than 5 feet wide
c. more than 5 feet deep
d. deeper than they are wide

_______ 4. According to scaffolding accident statistics, which of the following categories has had the highest percentage of accidents?

a. Carpenters using scaffolds
b. Scaffolds under 10 feet tall
c. Injuries were a result of a fall
d. Lack of training in use of scaffolds

_______ 5. OSHA regulations require the use of _____ on all supported or ground-based scaffolds to transfer the load to the supporting surface.

a. baseplates
b. end frames
c. cross braces
d. screwjack legs

_______ 6. To level an end frame while erecting a frame scaffold, _____ of the screwjack must be inserted into the scaffold leg.

a. no less than half
b. no more than one-quarter
c. at least one-third
d. a minimum of two-thirds

_______ 7. Visual inspection of scaffold parts should take place _____ times.

a. at least five
b. two to three
c. two to four
d. three

_______ 8. The surface on which a mobile tower rolls must be _____ level.

a. perfectly
b. within 3 degrees of
c. up to 5 degrees within
d. no more than 10 degrees out of

_______ 9. _____ are metal brackets installed on ladders to hold scaffold plank.

a. Goosers
b. Base plates
c. Mud sills
d. Ladder jacks

_______ 10. For job-built ladders, cleats should be uniformly spaced at _____.

a. 12 inches top to top
b. 10 inches bottom to top
c. 8 inches top to top
d. 10 inches bottom to bottom

Terms

Read the descriptions of jobsite equipment. Write the name of the correct piece of equipment on the corresponding blanks.

_______________ 1. Uses rolling tracks; carves layers of soil and rock pushing them into piles or thicker layers

_______________ 2. Has many hydraulic pistons to move the arm as it digs

_______________ 3. Highly maneuverable and can spin a full circle inside the space it takes up on the ground

_______________ 4. Can move pallets of material from delivery trucks to where it will be installed by workers

Put in Order

Put in order the following steps for erecting a scaffold. Write the corresponding numbers on the blanks.

_______ A. Place planks on top of the end frames. A cleat should be nailed to both ends of wood planks to prevent plank movement. Platform laps must be at least 12 inches, and all platforms must be secured from movement.

_______ B. Lay out the location of baseplates and screwjacks on mud sills. The end frames must be properly spaced for the guardrails and cross braces to be properly installed.

_______ C. Stand one of the end frames up and attach the cross braces to each side, making sure the correct length cross braces have been selected for the job. Connect the other end of the braces to the second end frame.

_______ D. The second level of frames may be hung temporarily over the ends of the first frames and then installed onto the coupling pins of the first-level frames. Special care must be taken to ensure proper footing and balance when lifting and placing frames.

_______ E. Use a level to plumb and level each frame. Adjust screwjacks or cribbing to level the scaffold.

_______ F. Install uplift protection pins through the legs and coupling pins. Wind, side brackets, and hoist arms can cause uplift, so it is a good practice to pin all scaffold legs together.

_______ G. As each frame is added, keep the scaffold bays square with each other. Repeat this procedure until the first horizontal scaffold run is erected.

Identify

Identify the following equipment found at a jobsite. Write the correct name on the line next to the item.

1. ________________________________

2. ______________________________

3. ______________________________

4. ______________________________

5. ______________________________

List

List four responsibilities, according to OSHA guidelines, of *employers* at a jobsite. Write the correct answers on the corresponding lines.

1. ______________________________
2. ______________________________
3. ______________________________
4. ______________________________

List two responsibilities, according to OSHA guidelines, of *employees* at a jobsite. Write the correct answers on the corresponding lines.

1. ______________________________

2. ______________________________

List five rules for a clean and safe working environment at a jobsite. Write the correct answers on the corresponding lines.

1. ______________________________

2. ______________________________

3. ______________________________

4. ______________________________

5. ______________________________

List eight potential problems to look for during a scaffold inspection. Write the correct answers on the corresponding lines.

1. ______________________________

2. ______________________________

3. ______________________________

4. ______________________________

5. ______________________________

6. ______________________________

7. ______________________________

8. ______________________________

Procedures

Read the following procedures. Determine what procedure is being described and write the name of the procedure on the blank.

Procedure 1: ________________________________

- Temporarily fasten rails of ladder side by side and layout dadoes 12" OC.
- Cut dadoes, cutting only ⅜" deep, any deeper will weaken the rail.
- Cut rungs 15–20" long. Fasten rungs in dadoes keeping ends of rungs flush with outside face of rails.

Procedure 2: ________________________________

- Cut 2 × 6 sawhorse top 36" long and bevel both edges of each end.
- If desired, bevel may be ripped on both edges for entire length of top.
- Cut four legs to 24" length or as desired, with bevel on each end at the same angle as top.
- Fasten all four legs to sawhorse top.
- Hold plywood brace as shown and mark its length at the top edge.
- From each mark, lay out same angle as top and legs.
- Cut, make duplicate, and fasten one on each end of horse flush with outside face of legs.

CHAPTER 7

Building Plans and Codes

Matching

Match terms to their definitions. Write the corresponding letters on the blanks. Not all terms will be used.

_______ 1. drawing in which the height of the structure or object is shown

_______ 2. close-up view of a plan or section

_______ 3. drawing showing a vertical cut-view through an object or part of an object

_______ 4. drawing in which three surfaces of an object are seen in one view

_______ 5. in an architectural drawing, an object as viewed from above

_______ 6. a multiview drawing

a. detail

b. elevation

c. isometric

d. orthographic

e. plan

f. section

g. horizon

h. dimension

True/False

Write *True* or *False* before the statement.

____________ 1. When put together, plans, elevations, sections, and details are called a set of prints.

____________ 2. Presentation drawings should not be colored.

____________ 3. Detail views are cutaway views.

____________ 4. The most common interior elevations are kitchen and bathroom wall elevations.

____________ 5. A finish schedule gives information about sizes and locations of windows and doors to be installed.

____________ 6. For complex commercial projects, a specifications guide has been developed by the Construction Specifications Institute (CSI).

____________ 7. The most commonly used scale found on prints is ¼ inch equals 1 foot.

____________ 8. Inspections ensure that construction is proceeding according to plan.

____________ 9. A setback inspection is for items such as electric and gas systems.

____________ 10. It is the contractor's responsibility to notify the building official when the construction is ready for a scheduled inspection.

Multiple Choice

Choose the best answer. Write the corresponding letter on the blank.

_____ 1. The lines in a(n) _____ view diminish the size as they converge toward vanishing points on a line called the horizon.

a. to-scale
b. perspective
c. isometric
d. outlook

_____ 2. _____ are blown up, with zoomed-in views of various items to show a closer view.

a. Elevations
b. Plan views
c. Presentations
d. Details

_____ 3. _____ are small parts drawn at a very large scale.

a. Details
b. Elevations
c. Plan views
d. Pictorials

_____ 4. Window _____ are printed instructions that give information about the location, size, and kind of windows to be installed in the building.

a. finishes
b. elevations
c. schedules
d. plots

_____ 5. _____ are written to give information that cannot be provided in the drawings or schedules.

a. Specs
b. Blueprints
c. Itemizations
d. Timetables

_____ 6. Lines that outline the object being viewed are called the _____ lines.

a. dimension
b. center
c. hidden
d. object

_____ 7. A _____ line is used in a drawing to terminate part of an object that, in actuality, continues.

a. leader
b. break
c. cutting-plane
d. section

_____ 8. Distances determined by zoning, such as maximum height and minimum lot width, are called _____.

a. differentials
b. zone types
c. setbacks
d. variances

_____ 9. Buildings and businesses that are not in their proper zones are called _____.

a. nonconforming
b. outliers
c. randoms
d. variances

_____ 10. A building permit is needed _____.

a. before the construction is complete
b. within 90 days of the start of construction
c. before construction can begin
d. only if construction involves special hazards

Terms

Read the descriptions of the following plans. Write the name of the plan on the corresponding blanks.

_______________ 1. Shows information about the lot; shows elevation heights and direction of the sloping ground; simulates a view looking down from a considerable height

_______________ 2. Shows a horizontal cut through the foundation walls; shows the shape and dimensions of the foundation walls and footings

_______________ 3. A view of the horizontal cut made about 4 to 5 feet above the floor; shows locations of walls, windows, doors, and fixtures

_______________ 4. Not always used; when used, they show support beams and girders as well as the size, direction, and spacing of these members

Read the descriptions of the following plan views. Write the name of the view on the corresponding blanks.

_______________ 5. Shows the building as seen from the street; can show front, back, right side, and left side

_______________ 6. Shows a cross section as if the building were sliced open to reveal its skeleton

_______________ 7. Shows the building from above, looking down

Identify

Identify the following types of architectural views and plans. Write the correct name on the line next to the item.

1. ______________________________

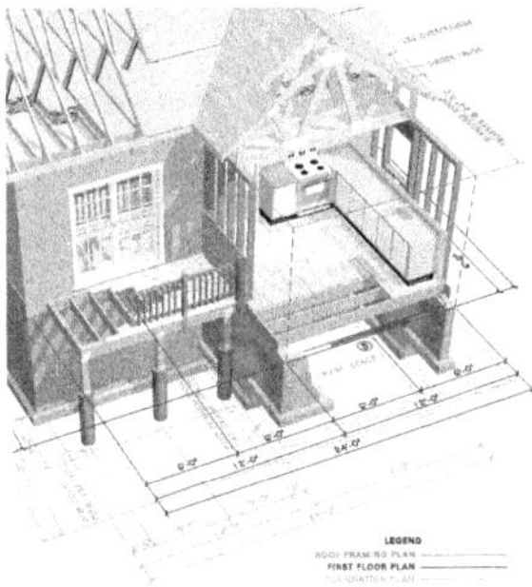

Courtesy of David Hultenius, Structural Engineer.

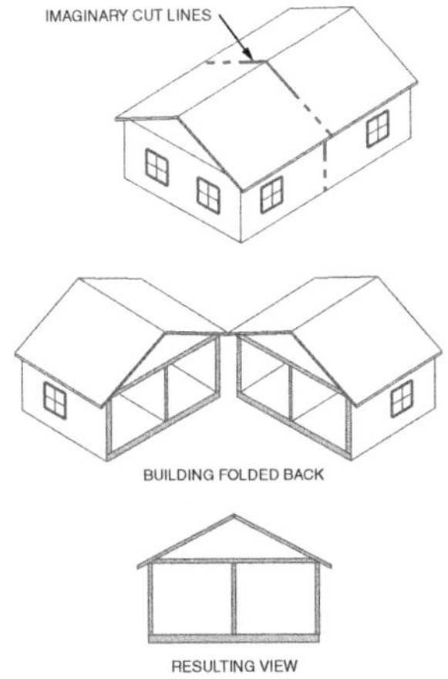

2. ______________________________

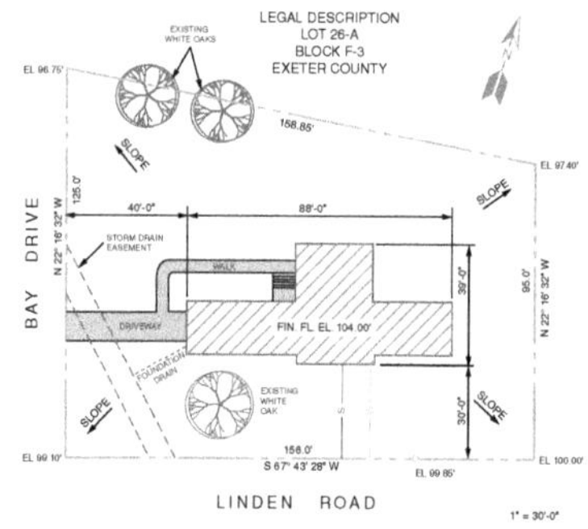

3. ______________________________

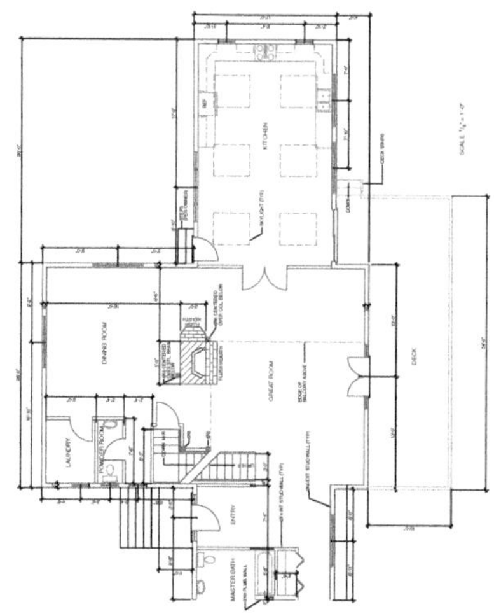

Courtesy of David Hultenius, Structural Engineer.

4. ______________________________

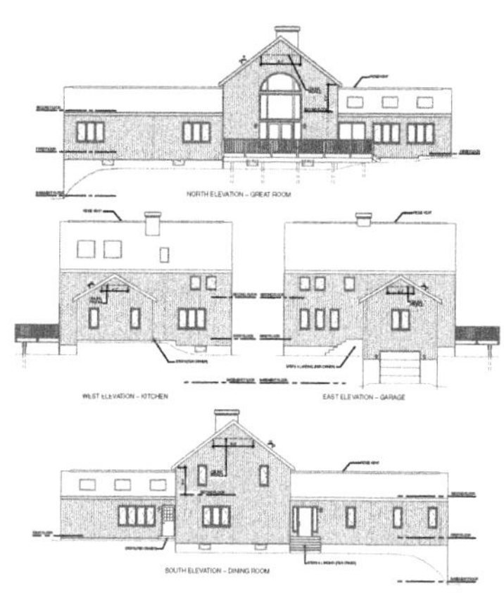

Courtesy of David Hultenius, Structural Engineer.

5. ______________________________

List

List four code inspections that a building inspector is likely to perform when visiting a jobsite. Write the correct answers on the lines below.

1. ______________________________

2. ______________________________

3. ______________________________

4. ______________________________

Activities

The lines below are commonly used in drawings. Write the type of line depicted on the corresponding blank.

A. ________________________

B. ________________________

C. ________________________

D. ________________________

E. ________________________

CHAPTER 8

Building Layout

Matching

Match terms to their definitions. Write the corresponding letters on the blanks. Not all terms will be used.

_______ 1. a telescope to which a spirit level is mounted

_______ 2. the part of a wall on which the major portion of the structure is erected

_______ 3. uses gravity to maintain a true level line of sight

_______ 4. a temporary or permanent supporting member for joists or other members running at right angles

_______ 5. optical leveling and plumbing instrument used in building construction

_______ 6. mathematical expression stating that the sum of the square of the two sides of a right triangle equals the square of the diagonal side

a. base
b. builder's level
c. compensator
d. foundation
e. laser
f. ledger
g. straightedge
h. Pythagorean theorem

True/False

Write *True* or *False* before the statement.

____________ 1. If no other tools are available, a carpenter's hand level and a long straightedge may be used together to level across a building area.

____________ 2. A builder's level has telescope that can be tilted up and down 45 degrees in each direction.

____________ 3. Excessive pressure on the leveling screws of an optical level may damage the threads of the base plate, resulting in errors.

____________ 4. In the spirit level, the bubble will always move in the same direction as your left thumb is moving when it rotates a leveling screw.

____________ 5. To sight an object, rotate the telescope and sight over its top, aiming it a few inches above the object.

____________ 6. If an optical level lens is dirty, rub it with your shirt tail or other available cloth.

____________ 7. With a little common sense, any employee on the jobsite can set up and operate laser instruments.

____________ 8. Locating the building is the practice of placing a building in the proper position on a building site.

Multiple Choice

Choose the best answer. Write the corresponding letter on the blank.

_____ 1. The laser beam of a laser level rotates _____ degrees, creating a level plane of light.

a. 360
b. 270
c. 180
d. 90

_____ 2. Using the 3-4-5 methods to place the third stake on the rear of the building involves _____.

a. using multiples of 3-4-5 to create larger right triangles
b. determining placement using the Pythagorean theorem
c. measuring the building length from stake two and the width from stake one
d. using measurements to the powers of 3, 4, and 5 to create isosceles triangles

_____ 3. _____ are wood frames built behind the layout stakes to which building layout lines are secured.

a. Ledger slats
b. Foundation walls
c. Batter boards
d. Layout panels

_____ 4. Drive the batter board stakes back from the building stakes _____.

a. four to six feet
b. no less than ten feet
c. between three and ten feet
d. at least equal to the depth of the excavation

_____ 5. Batter boards should be braced if they are _____.

a. placed in clay soil
b. placed in soft soil
c. over 1'-8" high
d. over 2'-0" high

Terms

Read the description of the uses for particular tools. Write the name of the correct tool on the corresponding blanks.

_______________ 1. Used for leveling from one point to another; its accuracy is based on the fact that water seeks its own level

_______________ 2. Used with an optical level; used for long sightings because of its clearer graduations

_______________ 3. Developed for the construction industry to provide more accurate and efficient layout work; uses a device that releases a narrow beam of light that can be seen over long distances

_______________ 4. Installed before excavation to allow layout stakes to be reinstalled quickly

Put in Order

Put in order the following steps for using a stick for a target to measure at two locations. Write the corresponding numbers on the blanks.

_______ A. Stick is placed next to grade stake and moved up or down until mark is sighted.

_______ B. Bottom of stick is placed on a desired elevation.

_______ C. Telescope is turned.

_______ D. Grade stake is marked in line with bottom of stick.

_______ E. Mark the stick.

Identify

Identify the following instruments used for carpentry. Write the correct name on the line next to the item.

1. ______________________________

Courtesy of David White.

2. ______________________________

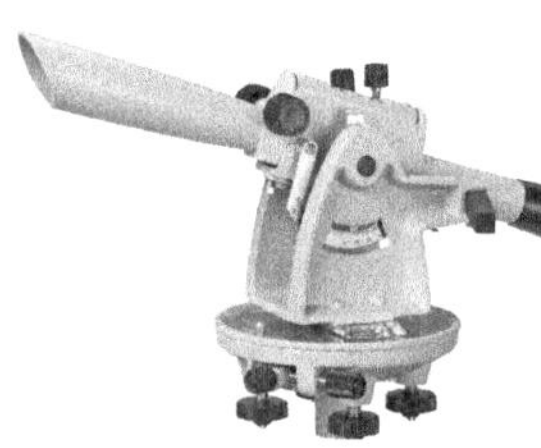

Courtesy of David White.

3. ______________________________

Courtesy of Laser Alignment, Inc.

4. ______________________________

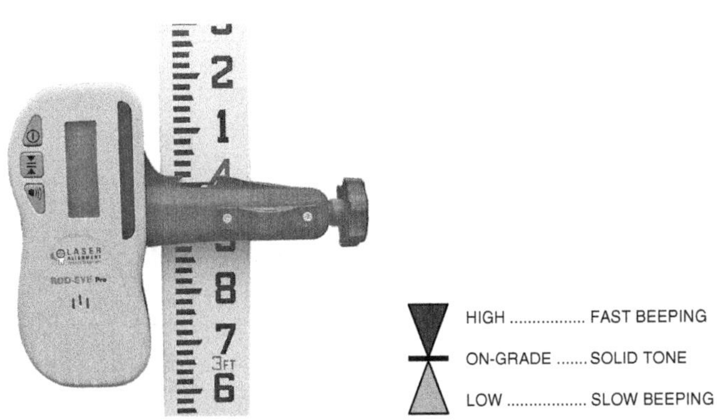

Courtesy of Laser Alignment, Inc.

List

List three safety precautions for using a laser. Write the correct answers on the lines below.

1. ______________________________

2. ______________________________

3. ______________________________

Procedures

Read the following procedure. Determine what procedure is being described and write the name of the procedure on the blank.

Procedure: ______________________

- EF is a straight line.
- Set up the transit level over point E and sight to point F.
- Tighten the horizontal clamp screw.
- Turn the horizontal circle scale to zero.
- Loosen the horizontal clamp screw and rate the telescope until 90 is read on the horizontal circle scale.
- Tighten the horizontal clamp screw.
- Drive a stake with a nail centered in its end to new sight point D.

Calculations

Use the information provided in the table to do your calculations. Write answers in the last column.

	Converting Feet-Inches to a Decimal Number		
	Feet-Inches		Decimal Feet
1.	40'-3"		
2.	15'-6"		
3.	27'-8"		
	Converting Decimal Number to Feet-Inches		
	Decimal Number		Feet-Inches to the Nearest 1/16
4.	36.28		
5.	27.57		
6.	18.08		
	Finding the Diagonal of a Rectangle Using the Pythagorean Theorem		
	Width	Length	Diagonal (ft-in to nearest 1/16)
7.	14'	30'	
8.	24'	60'	
9.	33'	92'	

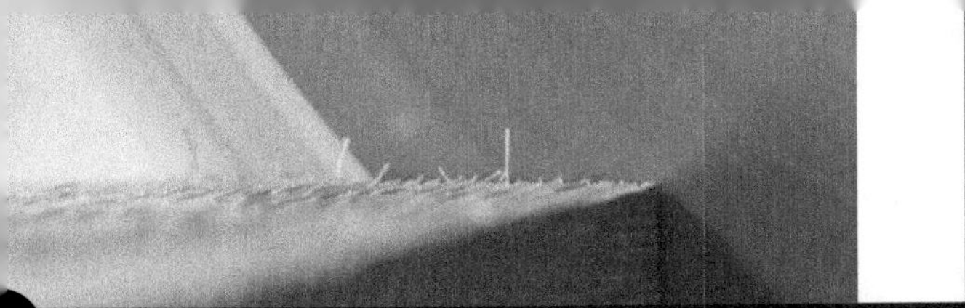

CHAPTER 9

Concrete Form Construction

Matching

Match terms to their definitions. Write the corresponding letters on the blanks. Not all terms will be used.

_______ 1. heavy timber or beam used to support vertical loads

_______ 2. the quality of being resistant to breaking under a pulling force

_______ 3. column built within and usually projecting from a wall to reinforce the wall

_______ 4. a rough frame used to form openings in poured concrete walls

_______ 5. vertical framing member in a wall running between plates

_______ 6. in stairs, the vertical distance of the flight; in roofs, the vertical distance from plate to ridge

_______ 7. a length of lumber or material applied over a joint to stiffen and strengthen it

_______ 8. a strip of wood used to keep other pieces a desired distance apart

_______ 9. the quality of being resistant to crushing

a. buck
b. compressive strength
c. girder
d. microsilica
e. pilaster
f. Portland cement
g. reinforcing rods
h. rise
i. run
j. scab
k. spreader
l. stud
m. tensile strength

True/False

Write *True* or *False* before the statement.

____________ 1. Footings are the first part of a building to be installed.

____________ 2. Building forms for slabs, walks, and driveways are exactly the same as building continuous footing forms.

____________ 3. A vapor retarder provides a barrier to soil gases, such as radon.

____________ 4. All wall form systems use similar components to make them perform as expected.

____________ 5. Outside corners of concrete-forming systems are created by attaching panels to angle irons with wedges.

____________ 6. Walers are difficult to install on forming systems.

____________ 7. Riser form boards for stairs are beveled on the top edge.

____________ 8. Water should be added to concrete so that it flows into forms without working it.

______ 9. A slump test shows the wetness or dryness of a concrete mix.

______ 10. Curing agents placed on green concrete contain harmful chemicals.

Multiple Choice

Choose the best answer. Write the corresponding letter on the blank.

______ 1. A combined slab foundation/footing is called a ____ slab.

a. dual-duty
b. monumental
c. colossal
d. monolithic

______ 2. When applying a vapor barrier of heavy plastic film under concrete, it should be lapped ____ and sealed.

a. no more than 2 inches
b. between 1 and 3 inches
c. approximately 3 inches
d. at least 4 inches

______ 3. ____ is the only insulation recommended for ground contact and can be placed between concrete and the subsoil.

a. Extruded polystyrene
b. Extruded aluminum
c. Polyethylene
d. Polypropylene

______ 4. Snap ties run through form boards and are wedged against additional form supports called ____.

a. form panels
b. anglers
c. walers
d. studs

______ 5. Concrete-forming system panels are tied together with ____ wedges.

a. formed plastic
b. metal
c. concrete
d. polyethylene

______ 6. The width of a step is called the ____.

a. rise
b. form
c. tread
d. edge

______ 7. Which of the following statements about the placement of concrete is true?

a. Working it by hand will cause voids or honeycombs.
b. Vibrators should not be used with a stiff mixture.
c. Overvibrating decreases the lateral pressure on a form.
d. Vibration allows trapped air to escape the concrete.

______ 8. Concrete must be protected from freezing for ____ after being placed.

a. at least four days
b. a minimum of ten days
c. no less than two weeks if the air is humid
d. between ten days and two weeks depending on humidity

______ 9. If moist-cured for seven days, concrete strength will be ____ full strength.

a. up to about 60 percent of
b. more than 75 percent of
c. kept to less than 40 percent of
d. nearly or completely at

Terms

Read the descriptions of terms associated with concrete construction. Write the correct term on the corresponding blanks.

________________ 1. Provide a base on which to spread the load of a structure over a wider area of the soil

________________ 2. Provides a lock between the footing and the foundation wall

________________ 3. Hold wall forms together at the desired distance apart; support both sides against lateral pressure of concrete

________________ 4. Used to reinforce concrete floor slabs resting on the ground, driveways, and walks

________________ 5. Performed by inspectors on a job to determine the consistency of the concrete

Put in Order

Put in order the following steps for constructing footing forms. Write the corresponding numbers on the blanks.

________ A. Fasten the spreaders across the form at intervals necessary to hold the form the correct distance apart.

________ B. Stretch lines back on the batter boards in line with the outside edge of the footing.

________ C. Erect the inside forms in a manner similar to the outside forms.

________ D. Drive corner stakes to the correct elevation and stretch lines between the stakes, if desired.

Identify

Identify the following items use with concrete construction. Write the correct name on the line next to the item.

1. ____

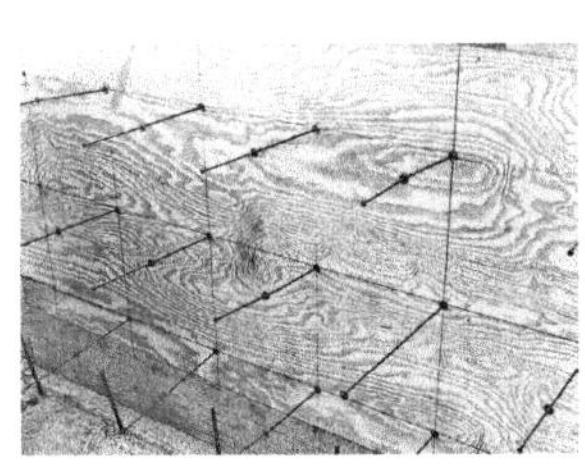

Courtesy of Dayton/Richmond Concrete Accessories.

2. ____

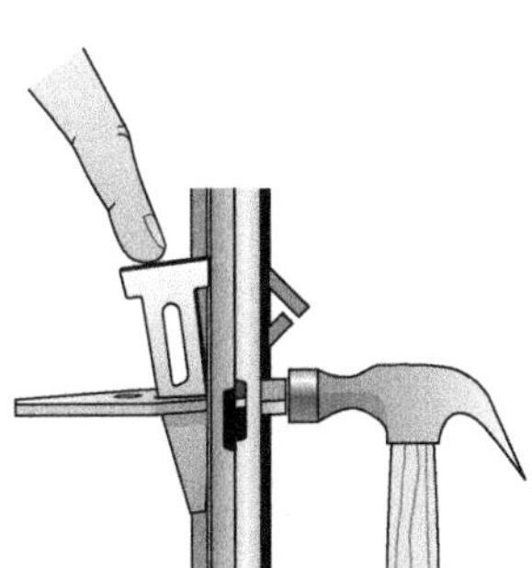

Courtesy of Dayton Superior Corporation.

3. ______________________________

4. ______________________________

Procedures

Read the following procedures. Determine what procedure is being described and write the name of the procedure on the blank.

Procedure 1: ______________________________

- Stretch lines again on the batter boards aligned with the outside of the foundation wall.
- Set panels directly on the concrete footing to the chalk line or on 2×4 or 2×6 lumber plates.
- Erect the outside wall forms first.
- Erect the panels for the inside of the wall.
- Install the walers. Let the snap ties come through them and wedge into place.
- Reinforce the corners with vertical 2×4s.
- If necessary, form the wall at intervals for the construction of pilasters, thickened portions of the wall.
- Brace the walls inside and outside as necessary to straighten them.
- The top surface of the concrete must be level and smooth to make attaching the wood frame easier.
- Set the anchor bolts into the fresh concrete.
- Install blockouts for larger openings.
- Install forms for girder pocket blockouts.

Procedure 2: ______________________________

- Determine the *rise* and *run* of each step and lay them out on the inside of the existing walls or form.
- Secure cleats (short strips of wood) to the side wall form at each riser location.
- For earth-supported stairs, place fill that is free of organic matter and compacted in place to reduce settling.
- For suspended stairs, construct a support made of plywood, joists, shores, and stair horses.
- Install the rebars or reinforcing as specified by the prints.
- Install riser boards that are ripped to width to correspond to the height of each riser.
- Brace the risers from top to bottom at mid-span.

Calculations

Use the information provided to complete your calculations. Write answers on the lines provided within the exercises.

Estimating the Amount of Concrete for a Slab

Estimate the cubic yards of concrete for the slab shown below. The depth of the slab is 6 inches. Use the space provided for your calculations.

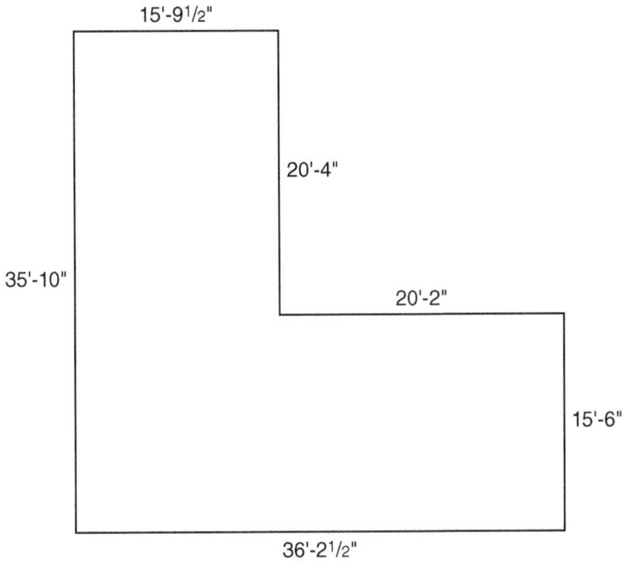

Total cubic yards: ______________________

Estimating the Amount of Concrete for a Footer

Estimate the cubic yards of concrete for a footer. Use the space provided for your calculations.

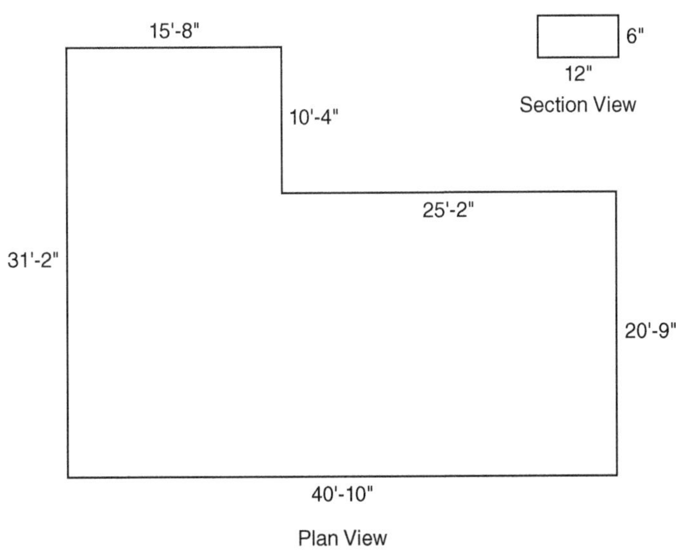

Total cubic yards: ______________________

Estimating Concrete Materials

Estimate the materials for a complete poured concrete foundation for a rectangular building 32' × 60'. Poured walls are 8 feet tall and 8 inches thick. Form boards are 12 feet long, and rebar is 20 feet long.

Estimate the materials for a foundation of a rectangular 30′ × 56′ building. Walls 8′ tall and 10″ thick, the footing is 2′ wide by 12″ thick, and slab is 5″ thick.

Item	Formula	Waste factor	Example
Footing form boards	footing PERM × 2 sides ÷ 12' = NUM of 12' boards		172' × 2 ÷ 12' = 28.6 ⇒ 29 boards
Slab form boards	slab PERM ÷ board LEN = NUM of boards		172' ÷ 12' = 14.3 ⇒ 15 boards
Wall forms	wall PERM × 2 sides ÷ 2' WID of form = NUM of form panels		172 × 2 ÷ 2' = 172 panels
Rebar footing	PERM × 2 ÷ 20' rebar LEN × waste = NUM of bars.	Add 10% for bar overlap	172' × 2 ÷ 20' × 1.10 =18.9 ⇒ 19 rebars
Rebar walls 2' × 2' grid horizontal	PERM × [wall HGT ÷ 2 FT grid − 1] ÷ 20' rebar LEN × waste = NUM of 20' HOR bars	Add 10% for bar overlap	[8' ÷ 2 − 1] × 172' ÷ 20 × 1.10 = 28.3 ⇒ 29 rebars
Rebar walls 2' × 2' grid vertical	PERM ÷ 2 ft grid + 1 per corner = NUM of VERT bars		172 ÷ 2 + 4 = 90 vertical rebars
Concrete footing (***Be sure all measurements are in terms of feet***)	footing width × footing depth × PERM ÷ 27 = CY	Add 5%	2' × 1' × 172' ÷ 27 × 1.05 5 5.6 1 5 3/4 CY
Concrete slab	slab WID × slab LEN × slab thickness ÷ 27 = CY concrete	Add 10%	30' × 56' × 5"/12 ÷ 27 × 1.10 = 28.5 ⇒ 28 3/4 CY
Concrete wall	PERM × wall HGT × wall thickness ÷ 27 = CY	Add 5%	172' × 8 × 10"/12 ÷ 27 × 1.05 = 44.5 ⇒ 44 3/4 CY

Footing form boards ____________________

Slab form boards ____________________

Wall forms ____________________

Rebar footing ____________________

Rebar walls 2 × 2 grid horizontal ____________________

Rebar walls 2 × 2 grid vertical ____________________

Concrete footings ____________________

Concrete slab ____________________

Concrete walls ____________________

CHAPTER 10

Floor Framing

Matching

Match terms to their definitions. Write the corresponding letters on the blanks. Not all terms will be used.

_______ 1. diagonal braces or solid wood blocks between floor joists used to distribute the load imposed on the floor

_______ 2. top or bottom horizontal member of a wall frame

_______ 3. thin, wedge-shaped piece of material used behind pieces to straighten them or for bringing their surfaces flush

_______ 4. a narrow board cut into studs of a balloon frame to support floor joists

_______ 5. a term used to describe when surfaces or edges are aligned with each other

_______ 6. boards or sheet material fastened to joists, rafters, and studs on which the finish material is applied

_______ 7. the member used to stiffen the ends of floor joists where they rest on the sill

_______ 8. long metal fastener with a threaded end used to secure materials to concrete

_______ 9. material used as the first floor layer on top of joists

a. anchor bolt
b. balloon frame
c. band joist
d. bridging
e. flush
f. joist
g. plate
h. ribbon
i. sheathing
j. shim
k. sill
l. subfloor

True/False

Write *True* or *False* before the statement.

_____________ 1. Lumber shrinks mostly across width and thickness and not so much along its length.

_____________ 2. In-line floor joists simplify installation of plywood subflooring.

_____________ 3. Floor joists rest on and transfer the load to the foundation.

_____________ 4. An anchor bolt should be located 18 inches from the ends of each sill.

_____________ 5. Engineered lumber should not be used in a floor system.

_____________ 6. Top edges of doubled joists should be flush or even.

_____________ 7. Cantilevered joists are supported at the inside end by a doubled joist or girder.

_____________ 8. Insects will not consume or destroy dry wood.

Multiple Choice

Choose the best answer. Write the corresponding letter on the blank.

_____ 1. _____ must be installed in the walls in several locations to prevent the spread of fire for a certain period of time.

a. Spacers
b. Stud plates
c. Draftstops
d. Insulation

_____ 2. For the 24-inch module method of framing, studs _____ with 24-inch spacing can be used in a single-story building.

a. up to 10 feet
b. up to 12 feet
c. greater than 12 feet
d. greater than 14 feet

_____ 3. In the _____, an energy-saving construction system, 2 × 6 wall studs are spaced 24 inches OC.

a. Oklahoma System
b. 2 × 6 Model
c. 24-inch Model
d. Arkansas System

_____ 4. If built-up girders of dimension lumber are used for floor frame members, _____ are fastened together.

a. at least five
b. no more than four
c. a minimum of three
d. a maximum of two

_____ 5. _____ lie directly on the foundation wall and provide a bearing for floor joists.

a. Headers
b. Sills
c. Box sealers
d. Rims

_____ 6. If joists are lapped over the girder, the minimum amount of overlap is _____, and the maximum overhang is _____.

a. 3; 12
b. 4; 10
c. 3; 10
d. 4; 12

_____ 7. Holes bored in joists for piping or wiring should _____.

a. not be located in the top one-third of the joist span
b. be located in the middle one-third of the joist span
c. not be closer than 2 inches from the top or bottom of the joist
d. be within 2 inches of the top or bottom of the joist

_____ 8. _____ row(s) of bridging is(are) needed for joists 8 to 16 feet long.

a. One
b. Two
c. Three
d. No

_____ 9. Wood steps should rest on a concrete base that extends _____ inches above the ground.

a. no more than 3
b. less than 5
c. between 2 and 5
d. at least 6

_____ 10. Termites generally will not eat _____.

a. hardwoods
b. treated lumber
c. lumber from evergreens
d. moist lumber

Terms

Read the descriptions of frame construction types. Write the correct construction type on the corresponding blanks.

________________ 1. Most commonly used in residential construction; the floor is built and then the walls are erected on top of it

________________ 2. The wall studs and first-floor joists rest on the sill; the second-floor joists rest on a ribbon that is cut flush with the inside edges of the studs

________________ 3. Uses fewer but larger pieces that are widely spaced

Put in Order

Put in order the following steps for installing girders. Write the corresponding numbers on the blanks.

______ A. Continue building and erecting sections until the girder is completed to the opposite pocket.

______ B. Sight the girder by eye from one end to the other and place wedges under the temporary supports to straighten the girder.

______ C. Set one end of the girder in the pocket in the foundation wall.

______ D. Check that temporary posts are strong enough to support the weight imposed on them until permanent ones are installed.

______ E. Place and fasten the other end on a braced temporary support.

______ F. Place permanent posts or columns after the girder has some weight imposed on it by the floor joists.

Identify

Identify the following types of frame construction. Write the correct name on the line next to the item.

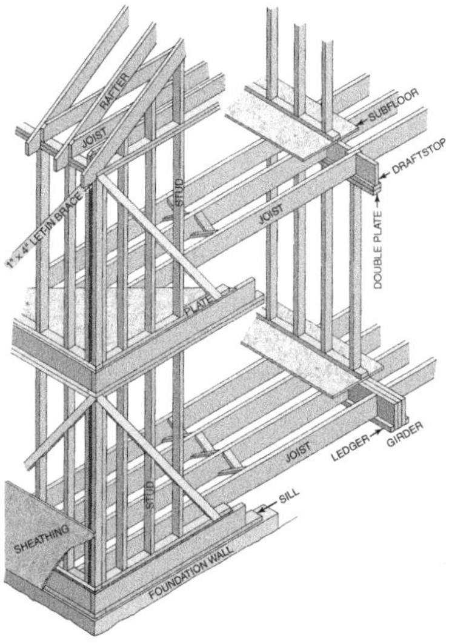

1. __________________________________

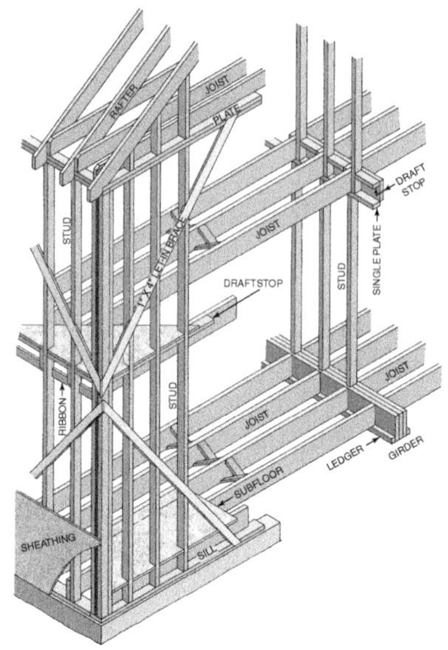

2. ______________________________

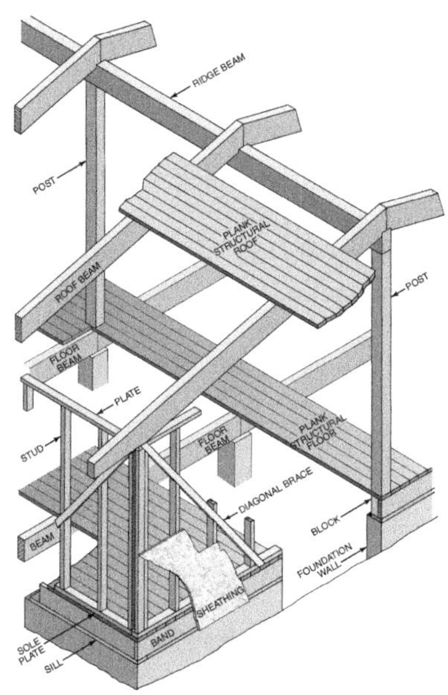

3. ______________________________

List

List two structural members for which building codes generally require the use of pressure-treated lumber. Write the correct answers on the corresponding lines.

1. ____

2. ____

Procedures

Read the following procedures. Determine what procedure is being described and write the name of the procedure on the blank.

Procedure 1: ________________________________

- Hold the sill in place against the anchor bolts to avoid covering the chalk line.
- Bore holes in the sill for each anchor bolt.
- Place the sill sections in position over the anchor bolts after installing the sill sealer.
- Replace the nuts and washers and level the sill by shimming where necessary.
- If the inside edge of the sill plate comes inside the girder pocket, notch the sill plate around the end of the girder.
- Raise the ends of the wood girder so it is flush with the top of the sill plate.
- Lower a steel girder for extra plate.

Procedure 2: ________________________________

- Fasten the inside trimmer joists in place.
- Mark the location of the header on the trimmers.
- Cut four headers to length by taking the measurement at the sill between the trimmers.
- Place two headers, one for each end of the opening, on the sill between the trimmers. Transfer the layout of the tail joists on the sill to the headers.
- Fasten the first header on each end of the opening in position by driving nails through the side of the trimmer into the ends of the headers.
- Fasten the tail joists in position by face nailing through the header and toenailing into girder or sill.
- Install inside header, face nailing into outside header and inside trimmer.
- Finally, double up the trimmer joists, and then add full-length joists to complete the frame.

Procedure 3: ________________________________

- Determine the actual distance between floor joists and the actual depth of the joist.
- Hold the framing square on the edge of a piece of bridging stock.
- Mark lines along the tongue and blade across the stock.
- Rotate the square, keeping the same face up. Align the same dimensions in the same fashion as before. Mark along the tongue.
- Make the actual length of the piece about ¼-inch shorter to ensure that it doesn't extend below the bottom edge of the joist.
- Bridging is then cut using a power miter box.

Procedure 4: ________________________________

- Check the girder for level, and adjust as needed.
- Measure accurately from the column footing to the bottom of the girder.
- Transfer this mark to the column. Deduct the thickness of the top and bottom bearing plates.
- To mark around the column so it has a square end, wrap a sheet of paper around it.
- Install the column's plumb breakline under the girder and centered on the footing.

- Fasten the top bearing plate to the girder with lag screws.
- If the girder is steel, the plates are then bolted or welded to the girder.
- Wood posts are installed in a similar manner, except their bottoms are placed on a pedestal footing.

Calculations

Use the information provided to complete your calculations. Write answers on the lines provided.

Estimating floor framing materials

Estimate the materials for a rectangular 28' × 48' building. A four-ply built-up full length girder, 16"OC joists lapped at the girder, 12' sill, band joists and girder material, four parallel partitions and one row of bridging each side of the girder.

Estimate the materials for a floor system of a rectangular 30′ × 56′ building. A two-ply LVL full length girder, 16″ OC joists lapped at the girder, 12′ sill and band joists, 5 parallel partitions and one row of bridging each side of girder.

Item	Formula	Waste factor	Example
Girder Built-up	girder LEN × NUM of plys ÷ LEN of each ply = NUM of boards		30′ × 2 ÷ 30′ = 2 LVL boards
Sill	PERM ÷ LEN of each piece = NUM of boards	Add steel girder sill	172′ ÷ 12 = 14.3 = 15 boards
Anchor Bolts	NUM Sills × 3 + one per corner = NUM of bolts		15 × 3 + 4 = 49 bolts
Joists-Band joist	building LEN × 2 sides ÷ length of each board = NUM of boards		56′ × 2 ÷ 12 = 9.3 = 10 boards
Joists-OC (On Center)	(building LEN ÷ OC spacing in ft + 1) × 2 if lapped + parallel partition supports = NUM joists		(56′ ÷ 16⁄12 + 1) × 2 + 5 = 91 joists
Bridging-Solid	building LEN × NUM of rows ÷ LEN joists = NUM of boards		56′ × 2 ÷ 16 = 7 boards
Bridging-Wood Cross	building LEN ÷ OC spacing in ft × ft per OC spacing) × NUM rows = lineal ft bridging		56′ ÷ 16⁄12 × 3 × 2 = 252 lineal ft bridging
Bridging-Metal	building LEN ÷ OC spacing in ft × 2 × NUM rows = NUM of pieces		56′ ÷ 16⁄12 × 2 × 2 = 168 NUM of pieces
Subfloor-sheets	building LEN ÷ 8 (ft per sheet) = round up to nearest ½ sheet = NUM pieces per row Building WID ÷ 4 (ft per sheet) = round up to nearest ½ sheet = NUM of rows NUM of pieces per row × NUM of rows = NUM of pieces		56′ ÷ 8 = 7 pieces per row 30′ ÷ 4 = 7.5 rows 7 × 7.5 = 52.5 = 53 pieces

Girder built-up ____________________

Sill ____________________

Anchor bolts ____________________

Band joist ____________________

Joists-OC ____________________

Bridging-solid ____________________

Bridging-wood cross ____________________

Bridging-metal ____________________

Subfloor sheets ____________________

CHAPTER 11

Wall and Ceiling Framing

Matching

Match terms to their definitions. Write the corresponding letters on the blanks. Not all terms will be used.

_______ 1. the underside trim member of a cornice or any such overhanging assembly

_______ 2. strips or blocks of wood installed in walls or ceilings used to fasten or support trim or fixtures

_______ 3. also called drywall; used to create a wall surface

_______ 4. pieces of dimension lumber installed between joist and studs

_______ 5. term for a structural member that carries weight from another part of the building

_______ 6. triangular-shaped section on the end of a building formed by the common rafters and the top plate line

a. backing

b. blocking

c. gable end

d. gypsum board

e. joist hanger

f. load-bearing

g. soffit

h. sole plate

True/False

Write *True* or *False* before the statement.

____________ 1. Corner posts are the same length as studs.

____________ 2. Wall bracing is required if rated panel wall sheathing is used.

____________ 3. All centerline dimensions for openings are measured from the building line, the outside edge of the exterior framing.

____________ 4. Sometimes the ceiling joists are installed in-line.

____________ 5. Wall furring should not be installed vertically.

____________ 6. The cores of SIPs can be made of expanded polystyrene or extruded polystyrene or urethane foam.

____________ 7. Wall studs are estimated from the total linear feet of exterior wall.

Multiple Choice

Choose the best answer. Write the corresponding letter on the blank.

_____ 1. Trimmers or jacks are fastened to full-length studs, often called _____ studs.

a. rough c. major
b. king d. main

_____ 2. In door openings, the trimmers, sometimes called _____, fit between the header and sole plate.

a. liners c. coating
b. layers d. films

_____ 3. A _____ is a heavily reinforced wall section designed to improve the lateral strength.

a. sheathing panel c. double-thick plate
b. shearwall d. curb

_____ 4. _____ is necessary between ceiling joists to support the top ends of partitions that run parallel to and between joists.

a. Support-type backing c. Ladder-type blocking
b. Weather backing d. Draftsop blocking

_____ 5. A _____ is used when extra support and stiffness is required on ceiling joists.

a. fullspan c. midspan
b. doubleback d. strongback

_____ 6. Ceiling joists should be toenailed into position with _____ nails.

a. no less than three 4-inch c. at least two 10d
b. no less than four 2-inch d. at least two 8d

_____ 7. The top and bottom plates of steel-framed wall are called _____.

a. rolled channels c. furring channels
b. plate studs d. runners or track

Terms

Read the descriptions of the parts of a wall frame. Write the name of the correct terms on the corresponding blanks.

_______________ 1. the top and bottom horizontal members of a wall frame

_______________ 2. vertical members of the wall frame that run the full length between plates

_______________ 3. form the top of wall openings and run at right angles to studs

_______________ 4. form the bottom of a window opening at right angles to the studs

_______________ 5. shortened studs that support the headers

_______________ 6. the same length as studs; historically made solid; today are constructed to allow for more insulation to penetrate the corners

_______________ 7. provides for fastening of interior wall covering in the corners formed by the intersecting walls

_______________ 8. horizontal members of the exterior wall frame in balloon construction; support the second-floor joists

Put in Order

Put in order the following steps for laying out a wall plate. Write the corresponding numbers on the blanks.

_______ A. Using straight lengths of lumber for the plates, cut, place, and tack two plates on the deck aligned with the chalk line.

_______ B. Measure in at the corners, on the subfloor, the thickness of the exterior wall. Snap chalk lines on the subfloor between the marks.

_______ C. Measure and mark the rough opening width.

_______ D. Mark the first on-center stud location by measuring in, from the outside corner, the regular stud spacing.

_______ E. Stretch a long tape the entire length of the wall section from the layout line of the first stud. Mark and square lines across the two plates for each stud.

_______ F. From the building prints, determine the centerline dimension of all the openings in the wall. Mark them on the sole plate by using a short line with a *C* over it.

_______ G. Build the wall section from the wall plate layout.

Identify

Identify the following structural components or activity related to wall and ceiling framing. Write the correct name on the line next to the item.

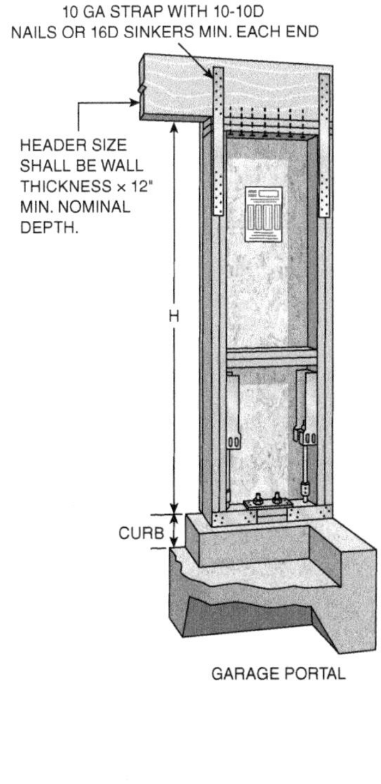

Courtesy of Simpson Strong-Tie Company.

1. ______________________________

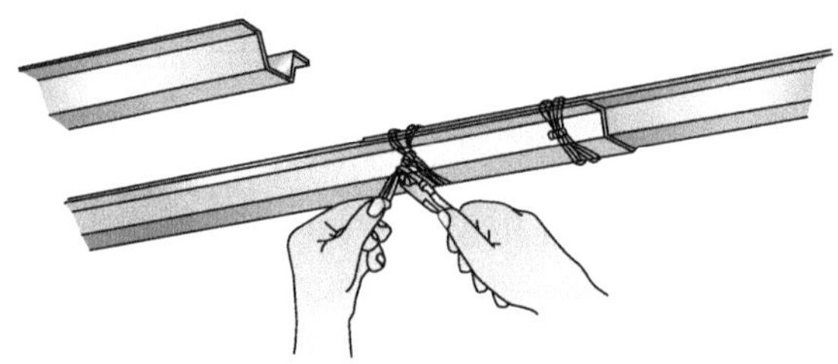

Courtesy of U.S Gypsum Corporation.

2. ______________________________

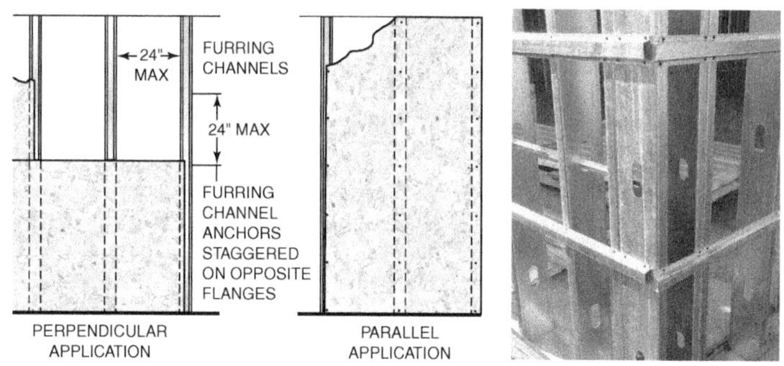

Courtesy of U.S Gypsum Corporation.

3. ______________________________

Courtesy of Insulspan.

4. ______________________________

Procedures

Read the following procedures. Determine what procedure is being described and write the name of the procedure on the blank.

Procedure 1: ______________________________

- To determine the correct measurement, add the following:
  - Sill thickness
  - Door height
  - Head jam thickness
  - Shim space between jamb and header

Procedure 2: ______________________________

- Pull the tack nails from the plates, turn up on edge, and separate them a distance equal to the stud length.
- Frame the openings first to make assembly easier.
- Place the jack studs, headers, and rough sills in position.
- Fasten the rough sills in position to the jack studs.
- Fasten a king or full-length stud to each jack stud by driving nails in a staggered fashion about 12 inches apart.
- Install the partition studs and then the remaining full studs.
- Install a doubled top plate.

- Recess or extend the doubled top plate to make a lap joint.
- Square the wall section by first aligning the bottom edge of the sole plate to the chalk line on the subfloor.
- Check square by measuring both diagonals from corner to corner.
- Install let-in bracing.

Procedure 3: ______________________________

- Remove the toenails from the top plate while leaving the toenails in the bottom plate.
- Lift the wall section into place, plumb, and temporarily brace.
- Check to be sure that the sole plate is still on the chalk line.
- Nail the sole plate to the band or floor joists below every 16 inches along the length.
- Fasten end studs in the corners together to complete the construction of the corner post.
- Straighten the wall by using a string with three blocks of equal thickness.
- Fasten a block to the side of each end of the top plate.
- Stretch a line tightly between the blocks so that these blocks hold the string off the wall.
- Use the third block as a gauge, adjusting for plumb.
- Adjust the wall in or out with each temporary brace until the gauge block just clears the line when held against the top plate.

Procedure 4: ______________________________

- Lay out spacing lines on the doubled top plate.
- For the first wall, mark an *R* for the rafter on one side of the layout line (which side of the line is not critical) and an *X* or a *C* on the other side.
- When joists are lapped, place the *R* of the second wall on the opposite side of the line as from the first wall.
- Mark the *X* and *C* accordingly.
- When joists are in-line, place the *R* on the same side of the line as the first wall and mark the *X* and *C*.
- The layout letters on the load-bearing partition must reflect the decisions made on the outside walls.

Activities

Identify various types of solid and built-up headers typically used in framing. Write the correct answers on the corresponding blanks.

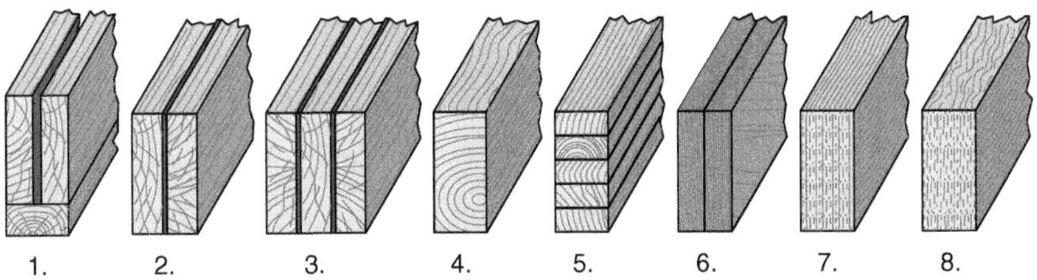

1. ______________________________

2. ______________________________

3. ______________________________

4. ______________________________

5. ______________________________

6. ______________________________

7. ______________________________

8. ______________________________

Label the components of a window opening. Write the correct answers next to the corresponding arrows.

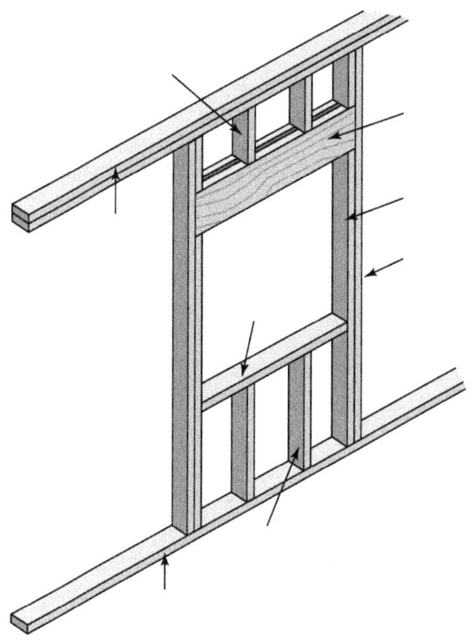

Label the components of a door opening. Write the correct answers next to the corresponding arrows.

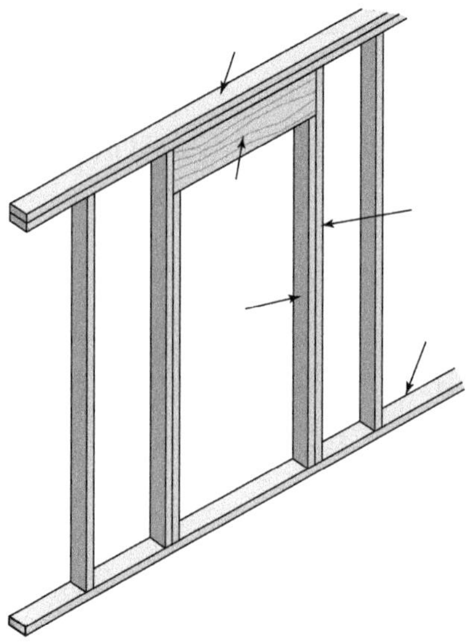

Calculations

Use the information provided to complete your calculations. Write answers on the lines provided.

Estimating exterior wall materials

Estimate materials for exterior walls of a rectangular building that measures 28' × 52'. The wall height is 8'-0", the box header is 1 foot wide, six 3646 windows and three 3'-0" × 6'-8" doors.

Estimate the materials for an exterior wall of a rectangular building that measures 30′ × 62′. The box header is one foot wide, six 3246 windows and three 3′-0″ × 6′-8″ doors.

Item	Formula	Waste factor	Example
Wall OC Studs	wall LEN in ft = NUM of studs		2 × (30 + 62) = 184′ = 184 PC
Wall plates	wall LEN × 3 ÷ 12′ = NUM 12′ plates	5%	184′ × 3 ÷ 12′ × 1.05 = 48.3 = 49 PC
Headers *win & door*	total of window & door WID plus 6″ per unit	5%	[6 (3′-2″ + 6″) + 3 (3′-0″ +6″)] × 1.05 = 32.5 = 33 LF
Rough sills	total of window WID plus 6″ per unit	5%	6 (3′-2″ + 6″) × 1.05 = 23.1 = 24 LF
Wall sheathing	wall area ÷ 32	5%	184′ × (8′ + 1′) ÷ 32 × 1.05 = 54.3 = 55 PC
Temporary braces	one per 20 feet of wall plus one per corner		184 ÷ 20 + 4 = 13.2 = 14 PC

____________________ Wall OC studs

____________________ 12' Wall plates

____________________ Headers (windows & doors)

____________________ Rough sills

____________________ Wall sheathing

____________________ Temporary braces

CHAPTER 12

Roof Framing

Matching

Match terms to their definitions. Write the corresponding letters on the blanks. Not all terms will be used.

_______ 1. common type of roof that pitches in two directions

_______ 2. horizontal framing pieces in a cornice, installed to provide fastening for the soffit

_______ 3. a direction to the side at about 90 degrees

_______ 4. vertical surface of the cornice finish installed on the bottom end of rafters

_______ 5. an inverted gable roof

_______ 6. the sloping portion of the gable ends of a building

_______ 7. intersection of two roof slopes forming an interior corner

_______ 8. a compound miter cut on the end of certain roof rafters

_______ 9. a roof that slopes up toward the ridge from all walls

_______10. type of roof that slopes in one direction only

_______11. a rafter that extends diagonally from the corner of the plate to the ridge at the intersection of two different roof surfaces

a. butterfly roof
b. cheek cut
c. dormer
d. fascia
e. gable roof
f. gambrel roof
g. hip jack rafter
h. hip rafter
i. hip roof
j. intersecting roof
k. lateral
l. lookout
m. rake
n. shed roof
o. valley

True/False

Write *True* or *False* before the statement.

_____________ 1. The gable roof is rarely used in modern buildings.

_____________ 2. Joints in the ridgeboard should be centered on a rafter; otherwise, a scab is needed.

_____________ 3. The rake rafter should not be straight along its side.

_____________ 4. The seat cut and tail are the same for the common rafter and the hip jack rafter.

_____________ 5. Hip and jack rafters are very similar.

_____________ 6. The shed roof rafter is similar to a common rafter but may require two seat cuts instead of one.

_____________ 7. The upper chords of a truss roof act as rafters.

Multiple Choice

Choose the best answer. Write the corresponding letter on the blank.

______ 1. A ______ has two different runs on either side of the ridge.

a. hip roof
b. shed roof
c. saltbox roof
d. dormer

______ 2. Because the hip rafter run is at a 45-degree angle from the plates, the amount of horizontal distance it covers is ______.

a. greater than that of the common rafter
b. less than that of the common rafter
c. the same as that of the common rafter
d. comparable to a valley rafter

______ 3. The ______ runs between a hip rafter and a valley rafter.

a. shortened valley rafter
b. valley cripple jack rafter
c. supporting valley rafter
d. hip-valley cripple jack rafter

______ 4. The total run of the supporting valley is the run of the common rafter of the main roof, called the ______.

a. ridge run
b. main run
c. major span
d. chief span

______ 5. Which of the following statements about roof trusses is correct?

a. They can support a roof over spans that can reach 100 feet.
b. They require load-bearing partitions below.
c. Framing takes more time than for gable roofs.
d. Because of their design, they have significant attic space.

______ 6. Nails in roof sheathing are spaced ______ inches apart on the ends and ______ inches apart on the intermediate supports.

a. 6; 18
b. 6; 12
c. 3; 18
d. 3; 12

Terms

Read the definitions of roof framing terms. Write the correct term on the corresponding blanks.

______________ 1. Sloping structural member of a roof frame; supports the roof covering

______________ 2. Horizontal distance covered by the roof; usually the width of the building measured from the outer faces of the frame

______________ 3. Horizontal distance over which the rafter covers; typically one-half the span

______________ 4. Total vertical distance that the roof rises; found by multiplying the unit rise by the total run of the rafter

_______________ 5. Horizontal member that forms the highest point of the roof system; secures the upper end of the rafters

_______________ 6. Forms the location where the rafter will sit on and be fastened to the wall

_______________ 7. Length of a rafter measured from the seat cut to the ridge

_______________ 8. The small right triangle found on the set of prints for the building

_______________ 9. Does not change from building to building; 12 inches for all common and jack rafters; 16.97 inches for valley rafters

_______________ 10. Distance that the roof rises vertically for every unit of run

_______________ 11. Length of rafter necessary to cover one unit of run

_______________ 12. Ratio of rise to span of a roof

_______________ 13. Refers to the steepness of a rafter; compares the unit rise to the unit run

_______________ 14. Any line of the rafter that is vertical when the rafter is in position

_______________ 15. Any line on the rafter that is horizontal when the rafter is in position

Procedures

Read the following procedures. Determine what procedure is being described and write the name of the procedure on the blank.

Procedure 1: ______________________________

- Lay a piece of rafter stock across two sawhorses.
- Begin at one end of the board. This will become the upper end or ridge plumb cut.
- Mark the rafter along the outside edge of the tongue. This is the first plumb line for the ridge.
- Shorten the rafter.
- Determine the rafter line length by first finding the unit length of the rafter from the top line of the rafter table.
- Make a plumb line for the seat cut the same way as for the ridge plumb line.
- Make a seat cut. On roofs with moderate slopes, the length of the level cut of the seat is often the width of the wall plate. For steep roofs, the level cut is shorter.

Procedure 2: ______________________________

- Lay the ridgeboard on top of the work platform in the same direction it was laid out.
- Fasten a rafter to each end of the first section of ridgeboard from the same side of the building.
- Raise the ridgeboard and two rafters into position.
- Fasten the rafters at the seat into the plate with three nails.
- Install two opposing rafters on the other side.
- Lift rafters into place and fasten the seat first to the plate. Then fasten the rafters to the ridgeboard.

- Drive fasteners of the second opposing rafter through ridge at a slight angle next to the first rafter.
- Plumb the section from the end of the ridgeboard to the outside edge of the plate at the end of the building. Brace the section temporarily from attic floor to ridge.
- Raise all other sections in a similar manner, installing the remaining rafters.
- Install collar ties as needed.

Procedure 3: ____

- Lay out the top plate for the location of each gable end stud.
- Using a level, plumb a stud from the layout mark on the plate. Mark along the top and bottom edge of the rafter the length of the stud. Measure this vertical distance.
- Determine the common difference in length of the studs. Multiply the OC spacing, in feet, times the unit rise.
- Lay out and cut the studs, longer and shorter as needed, from the first stud measured.
- Fasten the studs by toenailing to the plate and by nailing through the rafter into the stud.
- Sight the top edge of the end rafters for straightness periodically as gable studs are installed.
- After all gable studs are installed, the end ceiling joist is nailed to the inside edges of the studs.

Procedure 4: ____

- Start at the tail. Place the common rafter pattern on a piece of stock and mark the tail, seat cut, and seat cut plumb line.
- Mark the length of the rafter by measuring from the seat cut plumb line along the rafter edge. Draw a plumb line.
- Shorten the rafter for the thickness of the hip using the same procedure as for the hip shortening at the ridge.
- Lay out cheek cuts by measuring, again at right angles to the second plumb line, one-half the thickness of the jack rafter (3⁄4 inch for dimension lumber).
- Draw a third plumb line through this measurement.
- Next, draw a line over the top of the rafter from the third plumb through the midpoint of the squared line.
- This new line and the third plumb line become the cut lines.

Procedure 5: ____

- Select a straight length of stock for a pattern. Lay it across two sawhorses.
- Mark a plumb line at one end. Hold the tongue of the square at the unit rise and the blade of the square at 17 inches, the unit of run for the hip rafter.
- Shorten the rafter by measuring at right angles from the plumb line one-half the 45-degree thickness of the ridge.
- Lay out another plumb line through this measurement. From the top of this second plumb line, square a line over the top edge of the rafter.
- Mark the midpoint of that line.
- Lay out cheek cuts by measuring, again at right angles to the second plumb line, one-half the thickness of the valley rafter.

- Draw a third plumb line through this measurement.
- Next, draw a line over the top of the rafter from the third plumb through the midpoint of the squared line. This new line and the third plumb line become the cut lines of the ridge cut.
- Measure and mark, from the first plumb line drawn at the ridge end and along the edge of the rafter, the length of the rafter.
- Draw a plumb line through this mark.
- The valley seat cut has the same height above the seat cut as for the common rafter.
- Measure down, from the top of the common rafter, along the seat cut plumb line to the seat cut level line.
- Mark that same measurement on the valley, measuring down along the seat cut plumb line. Draw the level line.
- Lay out a deeper seat cut plumb line to allow for the thickness of the valley rafter.
- Measure back one-half the valley thickness (usually 3⁄4 inch) along the seat cut level line toward the tail of the rafter. Draw a new plumb line through this mark.

Activities

Label the components of the roof truss depicted below. Write the correct answers next to the corresponding arrows.

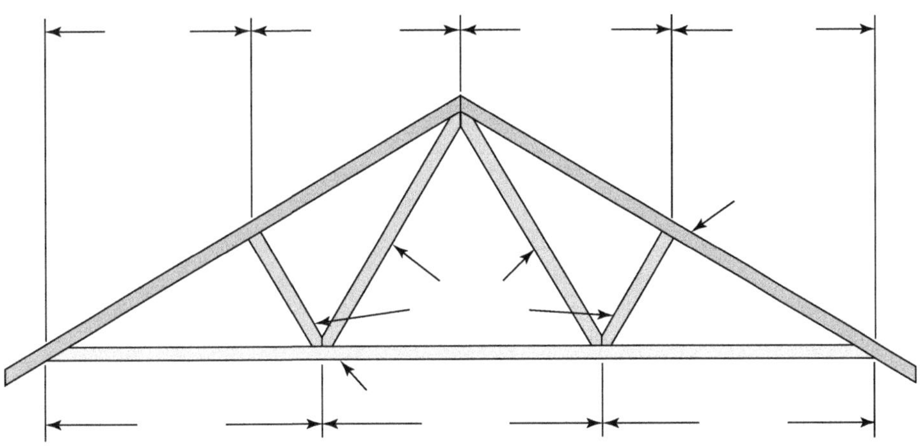

Calculations

Use the information provided to complete your calculations. Write answers on the lines provided.

Estimating roof frame materials

Estimate the materials to frame a roof of a rectangular 26' × 48' building. Framing 16" OC, 24" OC trusses with 16' temporary braces, 6" gable end overhang, a slope of 5 on 12 and a rafter length of 16'-11".

Estimate the materials for a roof of a rectangular 28' × 42' building. Framing 16" OC, 24" OC trusses, 9" gable end overhang, and a rafter length of 17'-4".			
Item	**Formula**	**Waste factor**	**Example**
Common Rafters	(building LEN (in ft) ÷ OC (ft) + 1) × 2 + 1 per each rake rafter.		42 ÷ 16⁄12 = 31.5 ⇒ 32 + 1 = 33 × 2 = 66 + 4 = 70 *PC*
Commons for Hip Roof	common rafters − rake rafters + 2 × NUM hips		(70 − 4) + 2 (4) 74 *PC*
Gable Ridgeboard	building LEN − (2 × gable end over hang)		42 + 2 × 9⁄12 = 435 *FT*
Hip Ridgeboard	building LEN − building WID + 4		42−28 + 4⁄12 + 14.333 *FT*
Gable end studs	building WID ÷ OC (ft) + 2		28 ÷ 16⁄12 + 21 + 2 = 23 *PC*
Trusses	building LEN ÷ OC (ft) − 1, plus two gable end trusses		42 ÷ 2 − 1 = 20 *PC*
Temp Truss Bracing	building WID ÷ 4 × building LEN ÷ 16 × 20% waste	20	28 ÷ 4 × 42 ÷ 16 × 1.20 = 222.05 ⇒ 23 *PC*
Gable Roof Sheathing	rafter LEN ÷ 4 (rounded up to nearest ½) × ridge LEN ÷ 8) × 2 (rounded up to nearest ½)		(17.333) ÷ 4.33 ⇒ 4.5)× (43.5 ÷ 8 = 5.4375 ⇒ 5.5) = 24.75 × 250 *PC*
Hip Roof Sheathing	gable sheathing × 5% waste	5%	50 × 1.05 = 52.5 ⇒ 53 *PC*
Gable End Sheathing	gable rise ÷ 4 (rounded up to nearest ½) × building WID ÷ 8 (rounded up to nearest ½)		14 × 7⁄12 = 8.166 ⇒ 8.5 × (28 ÷ 8 = 3.5 ⇒ 3.5) = 29.75 ⇒ 30 *PC*

1. Common rafters ________________________
2. Commons for hip roof ________________________
3. Gable ridgeboard ________________________
4. Hip ridgeboard ________________________
5. Gable end studs ________________________
6. Trusses ________________________
7. Temp truss bracing ________________________
8. Gable roof sheathing ________________________
9. Hip roof sheathing ________________________
10. Gable end sheathing ________________________

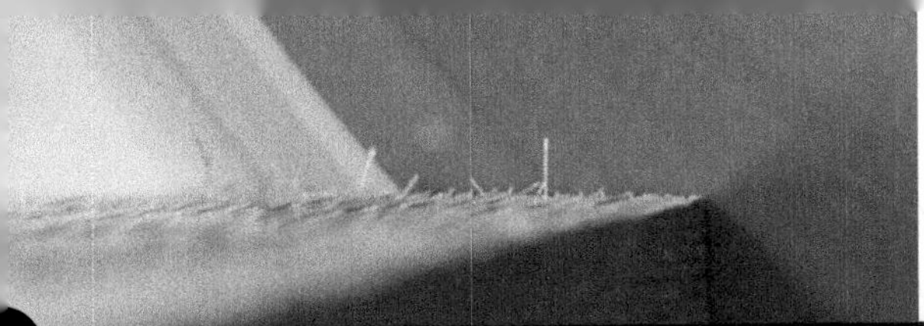

CHAPTER 13

Roofing

Matching

Match terms to their definitions. Write the corresponding letters on the blanks. Not all terms will be used.

_______ 1. flashing piece located on the lower side of a roof penetration such as a chimney, skylight, or dormer

_______ 2. accelerated oxidation of one metal because of contact with another metal in the presence of water

_______ 3. the unexposed part of roll roofing covered by the course above it

_______ 4. material used at intersections such as roof valleys and dormers and around window/door openings to prevent water entrance

_______ 5. metal edging strips placed on roof edges to provide support and overhang for the roofing material

_______ 6. a building paper saturated with asphalt used for waterproofing

_______ 7. a small, false roof built behind, or uphill from, a chimney or other roof obstacle; used to shed water

_______ 8. the amount that courses of siding or roofing are open to the weather

_______ 9. roof valley in which the roof covering meets in the center of the valley, completely covering the valley

a. apron
b. asphalt felt
c. closed valley
d. cricket
e. drip edge
f. electrolysis
g. exposure
h. flashing
i. mortar
j. open valley
k. selvage
l. square

True/False

Write *True* or *False* before the statement.

_____________ 1. Organic shingles have a base mat of glass fibers.

_____________ 2. Asphalt roofing materials are not flammable.

_____________ 3. Rake edge shingle cut pieces should be at least 4 inches in width.

_____________ 4. The starter course backs up and fills in the spaces between the butt seams of the first regular course of shingles.

_____________ 5. Roll roofing looks best on steep roofs.

_____________ 6. Flashing collars for plumbing stacks have a wide flange that rests on the roof deck.

_____________ 7. The flashing collar is fastened in place with one fastener in each upper corner.

_____ 8. Chimney flashings are installed in multiple overlapping layers.

_____ 9. Four nails are used for each piece of step flashing placed on the upper corner furthest away from the chimney.

_____ 10. Triangular wood pieces are cut and fit to create a cricket behind a chimney.

Multiple Choice

Choose the best answer. Write the corresponding letter on the blank.

_____ 1. A warning line is a barrier erected on a roof to warn employees they are approaching a distance of _____ from an unprotected roof side or edge.

a. 15 feet
b. 10 feet
c. 2 yards
d. 3 yards

_____ 2. One square of shingles will cover _____ square feet.

a. 10
b. 25
c. 50
d. 100

_____ 3. A metal drip edge is installed using nails spaced _____ inches along its inner edge.

a. 4 to 6
b. 6 to 8
c. 8 to 10
d. 10 to 12

_____ 4. When fastening asphalt shingles, use _____ fasteners in each strip of shingle.

a. at least five
b. no more than three
c. a minimum of four
d. between two and four

_____ 5. On steep pitch roofs, a _____ may be built on the upper or back side of the chimney to prevent the accumulation of water behind the chimney.

a. bridge
b. saddle
c. facade
d. conduit

_____ 6. _____ flashing is installed to cover the top of all step flashing and the apron flashing.

a. Cricket
b. Step
c. Offset
d. Counter

_____ 7. _____ is high-quality flashing material that is more expensive.

a. Copper
b. Aluminum
c. Galvanized steel
d. Mineral-surfaced asphalt roll roofing

_____ 8. An open valley should have a _____ -inch wide strip of flashing material centered in the valley.

a. 48
b. 36
c. 24
d. 12

_____ 9. _____ are individual metal pieces tucked between courses of shingles.

a. Woven valleys
b. Apron flashings
c. Step flashings
d. Flashing collars

Terms

Read the following roofing terms definitions. Write the correct terms on the corresponding blanks.

_______________ 1. Roof surface to which roofing materials are applied

_______________ 2. Number of overlapping layers of roofing and the degree of weather protection offered by roofing material

_______________ 3. Bottom exposed edge of a shingle

_______________ 4. Horizontal rows of shingles or roofing

_______________ 5. Height of the shingle or other roofing minus the exposure

_______________ 6. The vertical distance between courses of roofing; the amount of roofing in each course exposed to weather

_______________ 7. Horizontal distance that the ends of roofing in the same course overlap each other

_______________ 8. Adhesives to bond asphalt roofing products and flashings

Put in Order

Put in order the steps for installing asphalt shingles. Write the corresponding numbers on the blanks.

_______ A. Cut cap shingles and begin installation from one end.

_______ B. Install metal drip edge along the rake edges on top of the underlayment.

_______ C. Install the eaves drip edge, then apply underlayment.

_______ D. Prepare the starter course by cutting off the exposure tabs lengthwise through the shingle.

_______ E. Cut a set of starter shingles.

_______ F. Working horizontally across roof, add full shingles; start with the first course, second, and so on.

_______ G. Cover the two ridge fasteners with asphalt cement.

_______ H. Start the first course by fastening a full 36" shingle over the starter row at the rake edge.

_______ I. Apply the cap across the ridge until 3 or 4 feet from the end.

_______ J. Install the vented ridge cap as per the manufacturer's instructions, taking care to keep the vent ridge straight.

Procedures

Read the following procedures. Determine what procedure is being described and write the name of the procedure on the blank.

Procedure 1: ______________________________

- Apply first course of one roof, say the left one, into and past the center of the valley.
- Press the shingle tightly into the valley and nail, keeping the nails at least 6 inches away from the valley centerline.
- Cut shingles to adjust the butt ends so there is no butt seam within 12 inches of the valley centerline.
- Apply the first course of the other (right) roof in a similar manner, into and past the valley.
- Succeeding courses are applied by repeating this alternating pattern, first from one roof and then on the other.

Procedure 2: ______________________________

- Begin by shingling first roof completely, letting the end shingle of every course overlap the valley by at least 12 inches.
- Form the end shingle of each course snugly into the valley.
- Cut shingles to adjust the butt ends so there is no butt seam within 12 inches of the valley centerline.
- Snap a chalk line along the center of the valley on top of the shingles of the first roof.
- Apply the shingles of second roof, cutting the end shingle of each course to the chalk line.
- Place the cut end of each course that lies in the valley in a 3-inch wide bed of asphalt cement.

Procedure 3: ______________________________

- Snap a chalk line in the center of the valley on the valley underlayment.
- Apply the shingle starter course on both roofs.
- Trim the ends of each course that meet the chalk line.
- Fit and form the first piece of flashing to the valley on top of the starter strips.
- Trim the bottom edge flush with the drip edge.
- Fasten with two nails in the upper corners of the flashing only.
- Use nails of like material to the flashing to prevent electrolysis.
- Apply the first regular course of shingles to both roofs on each side of the valley, trimming the ends to the chalk line.
- Bed the ends in plastic asphalt cement.
- Do not drive nails through the metal flashing.
- Apply flashing to each succeeding course in this manner.

Calculations

Use the information provided to complete your calculations. Write answers on the lines provided.

Estimating roof frame materials

Estimate roofing materials for a roof with a 16'-6" gable rafter, 55' ridgeboard, and 5 bundles per square shingles.

Estimate the materials for a roof with a 15'-6" gable rafter, 54' ridgeboard, and 4 bundles per square shingles.			
Item	**Formula**	**Waste Factor**	**Example**
Roof Area	gable rafter LEN × gable ridgeboard LEN × 2 ÷ 100 SF per square = squares of roof area		15.5 × 54' × 2 ÷ 100 = 16.74 squares
Underlayment	roof squares ÷ 4 square per roll = rolls		16.74 ÷ 4 = 4.1 ⇒ 5 rolls
Asphalt Shingles	roof squares × bundles per square × waste factor = bundles	5%	16.74 × 4 × 105% = 70.3 ⇒ 71 bundles

1. Roof area ____________________
2. Underlayment ____________________
3. Asphalt shingles ____________________

CHAPTER 14

Windows and Doors

Matching

Match terms to their definitions. Write the corresponding letters on the blanks. Not all terms will be used.

_______ 1. molding used to trim around doors, windows, and other openings

_______ 2. vertical division between window units or panels in a door

_______ 3. the outside vertical members of a frame, such as a paneled door

_______ 4. protective plate covering the finished surface where penetrations occur, like door knobs

_______ 5. defect in lumber caused by a twist in the stock from one end to the other; twist in anything that should be flat

_______ 6. a coating on double-glazed windows; raises the insulting value by reflecting heat back into the room

_______ 7. a semicircular molding often used to cover a joint between two doors

_______ 8. decorative strips of wood used for finishing purposes

_______ 9. the horizontal member of a frame

_______10. slender strips of wood between lights of glass in windows in doors

_______11. the act of installing glass in a frame

_______12. that part of a window into which the glass is set

a. astragal
b. casing
c. deadbolt
d. escutcheon
e. glazing
f. housewrap
g. Low E
h. molding
i. mullion
j. muntin
k. rail
l. sash
m. stile
n. strike plate
o. wind

True/False

Write *True* or *False* before the statement.

_______________ 1. When shipped from the factory, the window is a complete unit except for the interior trim.

_______________ 2. Most windows are glazed with safety glass.

_______________ 3. The R-value can be decreased by using low emissivity glass.

_______________ 4. A double-hung window cannot be used in combination with other window types.

_______________ 5. Double doors are fitted and hung in a similar manner to single doors.

_______________ 6. Bifold doors may be installed in double sets, opening and closing from each side of the opening.

_______ 7. Codes require that doors swing inward in buildings used by the general public.

_______ 8. The passage lockset is used when locking a door is not necessary.

Multiple Choice

Choose the best answer. Write the corresponding letter on the blank.

_______ 1. Many windows come with false muntins, called _____ that do not actually separate or support the glass.

a. lights
b. rails
c. stiles
d. grilles

_______ 2. A(n) _____ jamb is cut to width to accommodate various wall thicknesses and applied to the inside edge of the jamb of a window frame.

a. vertical
b. traditional
c. extension
d. custom

_______ 3. Both sashes of a(n) _____ window slide vertically past each other in separate channels of the side jambs.

a. double-hung
b. fixed
c. casement
d. awning

_______ 4. A hopper window operates most similarly to the _____.

a. skylight
b. awning window
c. sliding window
d. single-hung

_______ 5. A(n) _____ is a molding that is rabbeted on both edge and designed to cover the joint between double doors.

a. astragal
b. rimming
c. wrap
d. track

_______ 6. _____ are narrow, closely spaced concave grooves that run parallel to the edge of trim.

a. Furring channels
b. Backbands
c. Spacers
d. Flutes

_______ 7. A _____ is the term for a twist in the door frame caused when the side jambs do not line up vertically with each other.

a. coil
b. curl
c. wind
d. twirl

_______ 8. The recess for a hinge is called a hinge _____.

a. leaf
b. gain
c. pin
d. loss

_______ 9. The _____ lockset has knobs on both sides that are turned to unlatch the door.

a. passage
b. privacy
c. deadbolt
d. limited-access

Terms

Read the following definitions for terms associated with windows and doors. Write the correct terms on the corresponding blanks.

________________ 1. When broken, the entire piece immediately disintegrates into a multitude of small granular pieces

________________ 2. Consists of two or sometimes three layers of glass separated by a sealed air space

________________ 3. Type of glazing with an invisible, thin, metallic coating bonded to the air space side of the inner glass

________________ 4. Vertical sides of the window frame

________________ 5. Piece of metal or plastic made slightly longer than the head casing; used to shed water over the top of the casing

________________ 6. Consists of two sashes; the upper one is fixed and the bottom one slides vertically

________________ 7. Consists of a sash hinged at the side; swings outward by means of a crank or lever

________________ 8. Consists of a frame in which a sash is hinged at the top; swings outward

________________ 9. Fixed in door frames on one or both sides of the door

________________ 10. Doors commonly used on wide clothes closet openings; installed on a double track; have rollers

________________ 11. Opened by sliding it sideways into the interior of a partition

Put in Order

Put in order the steps for installing a window. Write the corresponding numbers on the blanks.

_______ A. Level the window as needed with a temporary shim.

_______ B. Check the opening and window dimensions to verify the window will fit in the opening. Check the sill to verify it slopes slightly to the outside.

_______ C. Cut house wrap above the opening to allow housewrap to fold up.

_______ D. Install window flashing on rough framing.

_______ E. Install window flashing tape over the four nailing flanges.

_______ F. Caulk window flanges on the sides and top only.

_______ G. Fasten the lower end of both sides of the window.

_______ H. Place the window in the opening taking care to center the unit before bedding the unit in the caulk.

_______ I. Close and lock the sash, then plumb the sides of the unit.

_______ J. Fasten unit to the opening as per manufacturer's recommendations, and check that the sash operates properly.

Identify

Identify the following windows and doors. Write the correct name on the line next to the item.

Courtesy of Anderson Windows, Inc.

1. ______________________________

Courtesy of Anderson Windows, Inc.

2. ______________________________

Courtesy of Anderson Windows, Inc.

3. ______________________________

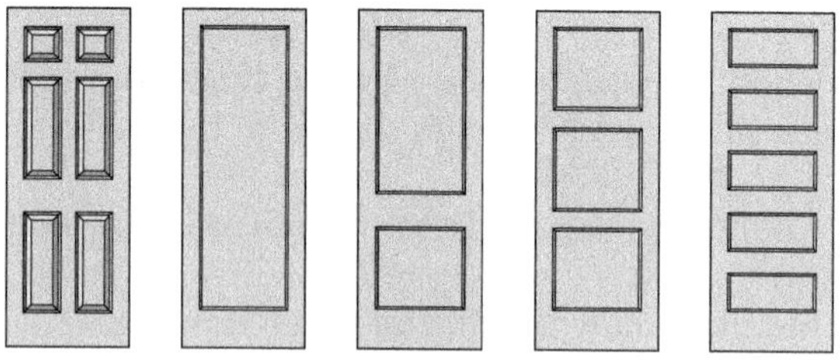

Courtesy of Morgan Manufacturing/Jeld-Wen, Inc.

4. ______________________________

Courtesy of Morgan Manufacturing/Jeld-Wen, Inc.

5. ______________________________

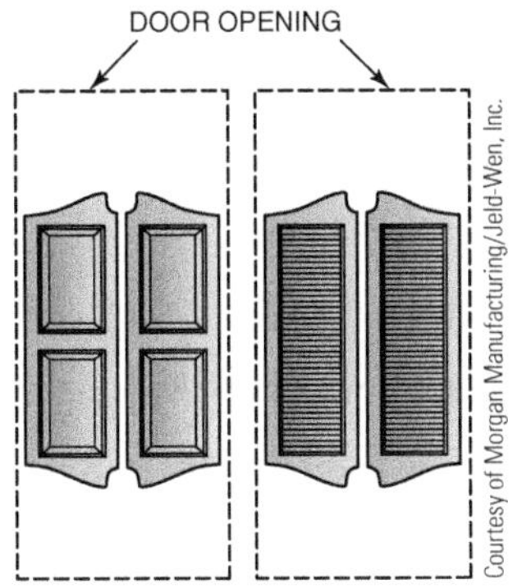
DOOR OPENING
Courtesy of Morgan Manufacturing/Jeld-Wen, Inc.

6. ______________________________

7. ______________________________

8. ______________________________

Courtesy of Morgan Manufacturing/Jeld-Wen, Inc.

9. ______________________________

Procedures

Read the following procedures. Determine what procedure is being described and write the name of the procedure on the blank.

Procedure 1: ______________________________

- Begin unrolling at a corner, place several fasteners in the upper corner.
- Unroll a short distance, 5–10 feet and align the top horizontal edge with the building keeping the sheet level.
- Place fasteners only along the upper edge.
- Repeat fastening along the top edge allowing the wrap to hang down the wall.
- Cut to length by allowing 12 inches to wrap the other wall or corner.
- Overlap all joints by at least 6 inches. Leave the cutouts for the opening until later.
- Move back to the center of the sheet, and smooth the wrap down vertically and fasten sheet every 8–12 inches.
- Move back to the center and then smooth out wrinkles diagonally toward the lower corners.
- Fasten as the sheet is it pulled smooth.
- Fasten the remainder of the sheet in a 12-inch grid or as directed by manufacturer.
- Cut out the openings by removing a rectangular piece that is smaller than the opening.
- Install other pieces making sure the horizontal joints are overlapped about 6 inches.

Procedure 2: ______________________________

- Set shim blocks under each jambs that are the same thickness as the finished floor.
- Center the unit in the opening so the door will swing in the desired direction.
- Be sure the door is closed and spacer shims are still in place between the jamb and door.
- Level the head jamb.
- Make adjustments by shimming the jamb that is low so it brings the head jamb level.
- Adjust a scriber to the thickness of shim and scribe this amount off of the other jamb.
- Remove frame and cut the jamb.
- Plumb the hinge side jamb of the door unit.
- Open the door and move to the other side.
- Check that the unit is nearly centered.
- Install shims between the side jambs and the rough opening at intermediate points, keeping side jambs straight.
- Nail through the side jambs and shims.
- Remove spacers from door edges.
- Check the operation of the door.

Procedure 3: ______________________________

- Attach rollers to the top of the door.
- Install pulls on the door.
- A special door pull contains edge and side pulls.

- Engage the rollers in the track by holding the bottom of the door outward in a way similar to that used with bypass doors.
- Test the operation of the door to make sure it slides easily and butts against the side jamb evenly.
- Make adjustments to the rollers, if necessary.
- Install the stops to the jambs on both sides of the door.

Procedure 4: ______________________________

- Check the contents and read the manufacturer's directions carefully. Mechanisms vary greatly.
- If the door is not pre-bored, open the door, and wedge the bottom to hold it in place.
- Measure up, from the floor, the recommended distance to the centerline.
- Position the center of the paper template on the squared lines.
- Lay out the centers of the holes that need to be bored.
- Bore the hole through the face of the door first.
- Bore the latch bolt hole through the edge of the door toward the center of the lockset hole.
- Use a faceplate marker to simplify lay out of the mortise for the latch faceplate.
- Complete installation according to the specific manufacturer's directions.
- Install the striker plate.

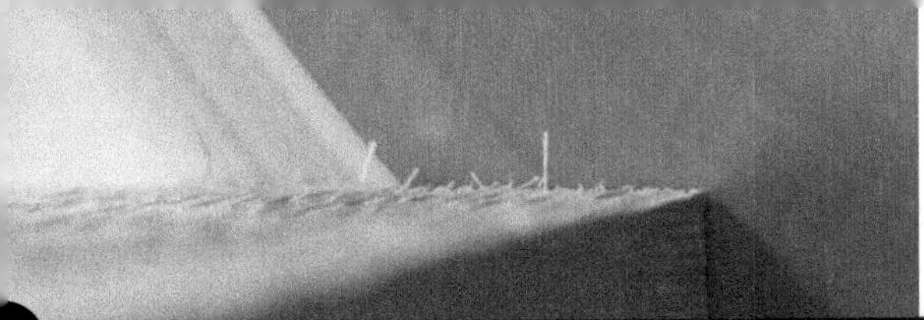

CHAPTER 15

Siding and Decks

Matching

Match terms to their definitions. Write the corresponding letters on the blanks. Not all terms will be used.

_______ 1. sloping portion of trim, such as on gable ends of a building

_______ 2. lower part of the roof that extends beyond and overhangs the sidewalls

_______ 3. vertical member of a stair rail; usually decorative

_______ 4. the finish member on the underside of a box cornice

_______ 5. part of an exterior finish that projects below another to cause water to drop off instead of running back against and down the wall

_______ 6. thin, narrow strip typically used to cover a joint in a vertical board

_______ 7. finish trim members used at the intersection of exterior walls

_______ 8. part of exterior trim applied to cover the joint between the cornice and the siding

_______ 9. method of fastening that conceals the fastener

_______ 10. finish material with random and finely spaced grooves running with the grain

_______ 11. a trough attached to an eave used to carry off water

_______ 12. tool used to bend sheet metal

_______ 13. narrow strip of wood used to lay out the installation heights of material such as siding or vertical members of a wall frame

_______ 14. the part of the exterior finish where the walls meet the roof

_______ 15. aluminum sheets sold in 50-foot long rolls and widths ranging from 12 to 24 inches

a. baluster
b. batten
c. blind nail
d. brake
e. coil stock
f. corner board
g. cornice
h. downspout
i. drip
j. eave
k. frieze
l. gutter
m. plancier
n. rake
o. story pole
p. striated

True/False

Write *True* or *False* before the statement.

_______ 1. Bevel siding is commonly known as clapboards.

_______ 2. The water table usually consists of a board and a drip cap installed around the perimeter.

_______ 3. Tongue-and-groove siding may be installed horizontally or vertically.

_______ 4. Panel siding should be installed vertically.

_______ 5. Wood shingles and shakes must be applied to sidewalls in double-layer courses.

_______ 6. Lookouts are decorative moldings used to cover the joints.

Multiple Choice

Choose the best answer. Write the corresponding letter on the blank.

_______ 1. When fastening strips of tongue-and-groove siding, strips should be blind nailed _____ inches apart.

a. exactly 18
b. approximately 12
c. from 16 to 24
d. no less than 18

_______ 2. When double-coursing shingle and shake siding, the starter course is _____.

a. quadrupled
b. tripled
c. doubled
d. a single layer

_______ 3. _____ are made with several opening sizes for vinyl siding panels.

a. C-grooves
b. J-channels
c. I-tunnels
d. L-furrows

_______ 4. The _____ is a horizontal framing member fastened to the rafter tails that provides an even, solid, and continuous surface for attachment of other cornice members.

a. subfascia
b. postfascia
c. rough soffit
d. false soffit

_______ 5. A downspout is also called a _____.

a. leader pipe
b. slip trough
c. gutter return
d. diverter

_______ 6. For decking run diagonally, joists should be spaced _____ inches OC.

a. 24
b. 20
c. 18
d. 16

_______ 7. Each linear foot of deck railing must be able to resist a pressure of _____ pounds per square foot applied horizontally at a right angle against the top rail.

a. 20
b. 25
c. 30
d. 50

Terms

Read the following descriptions of cornice designs. Write the correct cornice types on the corresponding lines.

_______________ 1. Helps protect the sidewalls from the weather; provides shade for windows; may be narrow or wide; may have level or sloping soffits

_______________ 2. Has no rafter overhang; frequently used on rakes of a gable end

_______________ 3. Has no soffit; installed directly to the underside of rafter tails

_______________ 4. Found on buildings with hip or mansard roofs; extends around the entire building

_______________ 5. Created by adding a soffit to an open cornice

Put in Order

Put in order the steps for installing horizontal siding. Write the corresponding numbers on the blanks.

_______ A. Install starter strip of the same thickness and width as the siding.

_______ B. Determine the exposure so that it is about equal both above and below the window sill.

_______ C. Make butt seams tight fitting between pieces.

_______ D. Fasten siding to each bearing stud or about every 16 inches.

_______ E. Snap lines at the top of each course of siding.

_______ F. Layout the desired exposures.

List

List ten components of a metal gutter system and provide a brief description for each on the lines provided below.

1. ____

2. ____

3. ____

4. ____

5. ____

6. ____

7. ____

8. ____

9. ____

10. ____

Procedures

Read the following procedures. Determine what procedure is being described and write the name of the procedure on the blank.

Procedure 1: ________________________________

- Slightly back-bevel the ripped edge.
- Place it vertically on the wall with the beveled edge flush with the corner similar to making a corner board.
- Face-nail the edge nearest the corner.
- Fasten a temporary piece on the other end of the wall to be sided, projecting below the sheathing by the same amount.
- Stretch a line to keep the bottom ends of other pieces in a straight line.
- Apply succeeding pieces by toenailing into the tongue edge of each piece.
- Make sure the edges between boards come up tight.
- Tack a strip in place where the last full strip will be located.
- Level from the top and bottom of the window casing to this piece of siding.
- Mark the piece.
- Use a scrap block of the material, about 6 inches long, with the tongue removed.
- Hold the block so its grooved edge is against the side casing and the other edge is on top of the tacked piece.
- Mark the vertical line on the siding by holding a pencil against the outer edge of the block while moving the block along the length of the side casing.
- Remove and cut the piece, following the layout lines carefully.
- Cut and fit another full strip in the same place as the previously marked piece.
- Fasten piece in position against window casing.

Procedure 2: ________________________________

- Install the first piece with the vertical edge plumb.
- Rip the sheet to size, putting the cut edge at the corner.
- The factory edge should fall on the center of a stud.
- Panels must also be installed with their bottom ends in a straight line.
- It is important that horizontal butt joints be offset and lapped, rabbeted, or flashed.
- Vertical joints are either shiplapped or covered with battens.
- Apply the remaining sheets in the first course in like manner.
- Cut around openings in a similar manner as with vertical tongue-and-grooved siding.
- Carefully fit and caulk around doors and windows.
- Trim the end of the last sheet flush with the corner.

Procedure 3: ________________________________

- Measure and lay out the width of the wall section for the siding pieces.
- Determine the width of the first and last piece.
- Cut the edge of the first panel nearest the corner.
- Install an undersill trim in the corner board or J-channel with a strip of furring or backing.

- Punch lugs along the cut edge of the panel at 6-inch intervals.
- Snap the panel into the undersill trim.
- Place the top nail at the top of the nail slot.
- Fasten the remaining nails in the center of the nail slots.
- Install the remaining full strips, making sure there is a ¼" gap at the top and bottom.

Calculations

Use the information provided to complete your calculations. Use the space provided to do your calculations.

Estimating siding materials

Estimate the siding needed to cover a rectangular building that measures:

- 28' × 46' with 9' high walls
- 7' gable height
- three 3949 windows
- four 4959 windows
- two 3' × 7' doors

Consider lineal feet:

- 8" log cabin siding
- squares of 16" wood shingles exposed to 6"
- number of squares of vinyl siding

Pattern	Nominal Width	Width: Dressed	Width: Exposed Face	Factor for Converting SF to Lineal Feet	Factor for Converting SF to Board Feet
Bevel and Bungalow	4	3½	2½	4.8	1.60
	6	5½	4½	2.67	1.33
	8	7¼	6¼	1.92	1.28
	10	9¼	8¼	1.45	1.21
Dolly Varden	4	3½	3	4.0	1.33
	6	5½	5	2.4	1.2
	8	7¼	6¾	1.78	1.19
	10	9¼	8¾	1.37	1.14
	12	11¼	10¾	1.12	1.12
Drop T&G and Channel Rustic	4	3⅜	3⅛	3.84	1.28
	6	5⅜	5⅛	2.34	1.17
	8	7⅛	6⅞	1.75	1.16
	10	9⅛	8⅞	1.35	1.13
Log Cabin	6	5 7/16	4 15/16	2.43	2.43
	8	7⅛	6⅝	1.81	2.42
	10	9⅛	8⅝	1.39	2.32
Boards	2	1½	The exposed face width will vary depending on size selected and on how the boards-and-battens or boards-on-boards are applied.		
	4	3½			
	6	5½			
	8	7¼			
	10	9¼			

Coverage Estimator

	Approximate Coverage of One Square (4-bundle roof-pack) of Shingles at Indicated Weather Exposures:												
Length	3½"	4"	4½"	5"	5½"	6"	6½"	7"	7½"	8"	8½"	9"	9½"
16"	70	80	90	100	110	120	130	140	150	160	170	180	190
18"	—	72½	81½	90½	100	109	118	127	136	145½	154½	163½	172½
24"	—	—	—	—	73½	80	86½	93	100	106½	113	120	126½

Length	10"	10½"	11"	11½"	12"	12½"	13"	13½"	14"	14½"	15"	15½"	16"
16"	200	210	220	230	240	—	—	—	—	—	—	—	—
18"	181½	191	200	209	218	227	236	245½	254½	—	—	—	—
24"	133	140	146½	153	160	166½	173	180	186½	193	200	206½	213

Estimate the siding needed to cover a rectangular building that measures 30'×56' with 9' high walls, 9' gable height, three 3949 windows, four 4959 windows, and two 3'×7' doors. Consider the lineal feet of 8" bevel siding, squares of 18-inch wood shingles exposed to 7½", and number of squares of vinyl siding.

Item	Formula	Waste factor	Example
Building Area	PERM × wall HGT + 2 × gable area – opening area 5 build area		172 × 9 + (2 × ½ × 30 × 9) – (3 × 3 × 4 + 4 × 4 × 5 + 2 × 3 × 7) = 1548 – 158 = 1390 SF
8" Bevel Siding	building area × conversion factor (Figure 15-40) × waste = lineal ft	10%	1390 × 1.92 × 110% = 2935.6 ⇒ 2936 lineal feet
18" Wood Shingles exposed to 7½"	building area ÷ conversion factor (Figure 15-41) × waste = squares of siding	10%	1390 ÷ 136 × 110% = 11.2 ⇒ 12 squares of siding
Vinyl Siding	building area ÷ 100 × waste = squares of siding	10%	1390 ÷ 100 × 110% = 15.2 ⇒ 16 squares of siding

Perimeter
Building area
8" Log cabin siding
16" Wood shingles exposed to 6"
Vinyl siding

CHAPTER 16

Insulation and Ventilation

Matching

Match terms to their definitions. Write the corresponding letters on the blanks. Not all terms will be used.

_______ 1. caused by vibration, such as from a dropped object or footsteps on a floor

_______ 2. an additional sash placed on the outside of a window to create a dead air space to prevent the loss of heat from the interior in cold weather

_______ 3. fiberboard ceiling tile whose surface consists of small holes and fissures that act as sound traps

_______ 4. temperature at which moisture begins to condense out of the air

_______ 5. narrow strips of thin metal or other material applied to windows and doors to prevent the infiltration of air and moisture

_______ 6. material used to restrict the passage of heat or sound

_______ 7. droplets of water that accumulate on a cool surface when water vapor cools and changes to a liquid

_______ 8. number given to material to indicate its resistance to the passage of heat

_______ 9. unwanted movement of air into an insulation layer or a conditioned space

a. acoustical tile
b. air filtration
c. condensation
d. dew point
e. equilibrium
f. impact noise
g. insulation
h. R-value
i. storm sash
j. vapor retarder
k. weather stripping

True/False

Write *True* or *False* before the statement.

_______________ 1. All materials used in construction have some insulating value.

_______________ 2. Heat energy is thought to move from warmer materials into cooler ones in an attempt to reach equilibrium.

_______________ 3. Insulation becomes more effective as the air spaces become larger and fewer in number.

_______________ 4. Insulation pays for itself in saved energy dollars by reducing energy needs and lessening the carbon footprint of a building.

_______________ 5. Insulating for maximum comfort does not necessarily provide maximum economy of heating and cooling operations.

_______________ 6. Fiberglass is made of 75 percent recycled materials.

_____ 7. Polyurethane foams have the best R-value per inch.

_____ 8. Alcohol can be used to remove foamed-in-place polyurethane products from skin.

_____ 9. Cold air can hold more moisture than warm air.

_____ 10. Insulation that absorbs water has a reduced thermal resistance.

_____ 11. As the STC number gets higher, the sound barrier of the wall, ceiling, or floor gets better.

_____ 12. An INR rates the transfer of noise from impact.

Multiple Choice

Choose the best answer. Write the corresponding letter on the blank.

_____ 1. _____ is an example of electromagnetic radiation.

a. Visible light
b. Sound
c. Vibration
d. Lightning

_____ 2. _____ insulation is manufactured in blanket and batt form.

a. Loose-fill
b. Flexible
c. Rigid
d. Reflective

_____ 3. _____ is an example of rigid insulation.

a. Aluminum foil
b. Cork
c. Polyurethane foam
d. Batt

_____ 4. Cellulose insulation is produced from _____ percent recycled newspaper.

a. approximately 25
b. about 50
c. 60 to 70
d. 100

_____ 5. Voids in insulation of _____ percent of the overall area can create an R-value efficiency reduction of _____ percent.

a. 1; 50
b. 1; 75
c. 5; 25
d. 5; 50

_____ 6. Polyethylene is a vapor retarder that comes in rolls of _____ feet in length and widths of _____ feet.

a. 20; up to 16
b. 10 to 50; either 10 or 12
c. up to 60; no more than 10
d. up to 100; up to 20

_____ 7. _____ percent of all moisture that travels through a wall or ceiling does so by air filtration.

a. More than 95
b. Less than 50
c. Up to 25
d. No more than 75

_____ 8. Which of the following STC numbers would be required for a barrier that would make loud speech audible as a murmur?

a. 42
b. 36
c. 28
d. 25

Terms

Using the descriptions below, decide which mode of heat transfer applies to the different methods of cooking popcorn. Write the correct term for the heat transfer method on the corresponding blanks.

________________ 1. Stove-top method: popcorn absorbs heat from the electric heating element

________________ 2. Microwave package: uses electromagnetic waves to heat the kernels

________________ 3. Hot-air poppers: heated air blows onto the kernels

Put in Order

Put in order the steps for installing flexible insulation. Write the corresponding numbers on the blanks.

_______ A. Fold over flanges of facing over framing member.

_______ B. Use a hand- or hammer-tacker stapler to fasten the insulation in place or push friction fit insulation into place.

_______ C. Install the positive ventilation chutes and air-insulation dam between rafters in line with or on the exterior sheathing.

_______ D. Use scrap wood to compress the insulation, and cut the material to required length in one pass.

_______ E. Fill any spaces around windows and doors with spray can foam.

_______ F. Place the batts or blankets between the framing members.

Identify

Identify the following types of insulation. Write the correct name on the line next to the item.

1. ______________________________

2. ______________________________

3. ______________________________

List

List three single wall types that have a higher STC rating than a ½" gypsum wall board. Write the answers on the corresponding lines.

1. ____

2. ____

3. ____

List three ceiling combinations with better STC ratings than ⅜" gypsum board with 2 × 8 joists. Answers can include options for 2 × 10 joists. Write the answers on the corresponding line.

1. ____

2. ____

3. ____

CHAPTER 17

Wall Finish

Matching

Match terms to their definitions. Write the corresponding letters on the blanks. Not all terms will be used.

_______ 1. the best appearing side of a piece of wood or the side that is exposed when installed

_______ 2. a wall finish applied partway up the wall from the floor

_______ 3. a building product made by compressing wood fibers into sheet form

_______ 4. an edge of lumber whose sharp corners have been rounded

_______ 5. a panel used as a finished surface material made from a mineral mined from the earth

a. backing

b. eased edge

c. face

d. gypsum board

e. hardboard

f. veneer

g. wainscoting

True/False

Write *True* or *False* before the statement.

____________ 1. Gypsum is the brand name for sheetrock.

____________ 2. Coreboards and liner boards come in 2-foot widths.

____________ 3. Adhesives are not used for bonding gypsum board directly to supports.

____________ 4. Supplemental fasteners should be used with contact adhesives.

____________ 5. Gypsum panels are applied first to ceilings and then to the walls.

____________ 6. The floating angle method of drywall application omits fasteners in the corner intersection of the ceilings and wall panels.

____________ 7. Drying type joint compounds are used when a faster setting time is desired.

____________ 8. A clinching tool is sometimes used to set corner beads to the proper angle.

Multiple Choice

Choose the best answer. Write the corresponding letter on the blank.

_____ 1. Type X gypsum board is commonly called _____.

a. backing board
b. fire-code board
c. blue board
d. eased edge

_____ 2. _____ is used for the core of solid gypsum partitions.

a. Predecorated
b. Veneer plastic
c. X-rock
d. Coreboard

_____ 3. _____ has a special fire-resistant core encased in a moisture-repellent paper.

a. Liner board
b. Blue board
c. Tapered gypsum
d. Regular gypsum panels

_____ 4. _____ has a core of Portland cement reinforced with a glass fiber mesh embedded in both sides.

a. Coreboard
b. Drywall
c. Wonder board
d. Sheetrock

_____ 5. For laminating gypsum boards to each other _____ adhesive is used.

a. contact
b. construction
c. drywall
d. drywall stud

_____ 6. When single-fastening gypsum panels, fasteners are spaced _____ inches OC on ceilings and 8 inches OC on walls.

a. a minimum of 12
b. approximately 8
c. a maximum of 7
d. no more than 6

_____ 7. Corner bead is used during drywall application to _____.

a. replace joint compound
b. prevent corner pieces from touching
c. make corner pieces plumb
d. reinforce and protect exterior corners

_____ 8. The fastest setting type joint compound will harden in _____.

a. as little as 20 to 30 minutes
b. no less than 45 minutes
c. approximately 4 hours
d. about 6 to 8 hours

_____ 9. Control joints are metal strips with flanges on both sides of a _____ -inch, V-shaped slot.

a. ⅛
b. ¼
c. ½
d. ¾

_____ 10. In cold weather, the interior temperature must be a _____ before and during application of joint compound and for at least 4 days after application has been completed.

a. maximum of 80 degrees F for 6 hours
b. minimum of 70 degrees F for 12 hours
c. maximum of 60 degrees F for 18 hours
d. minimum of 50 degrees F for 24 hours

Terms

Read the following descriptions of gypsum panels. Write the type of panel being described on the corresponding blanks.

________________ 1. Most commonly used type; applied to interior walls and ceilings

________________ 2. Has a special tapered, rounded edge; produces a much stronger concealed joint than a tapered, square edge

________________ 3. Has greater resistance to fire because of special additives

________________ 4. Chemically treated to repel moisture; frequently used in bathrooms

________________ 5. Designed to be used as a base layer in multilayer systems

________________ 6. Faced with a specially treated blue paper; designed to receive application of veneer plastic

Put in Order

Put in order the steps for cutting and fitting gypsum board. Write the corresponding numbers on the blanks.

________ A. Take measurements to within ½ inch for the ceiling and ¼ inch for the walls.

________ B. Score the backside paper leaving some material at the top and bottom to act as a hinge.

________ C. Guide the knife with a drywall T-square using your toe to hold the bottom.

________ D. Smooth ragged edges with a drywall rasp, coarse sanding block, or knife.

________ E. Bend the board back against the cut. The board will break along the cut face.

________ F. Lifting the panel off the floor, snap the cut piece back quickly to the straight position. This will complete the break.

________ G. Using a utility knife, cut the board by first scoring the face side through the paper to the core.

Identify

Identify the materials and tools (if one is being used) in the following images. Write the correct name(s) on the line next to the image.

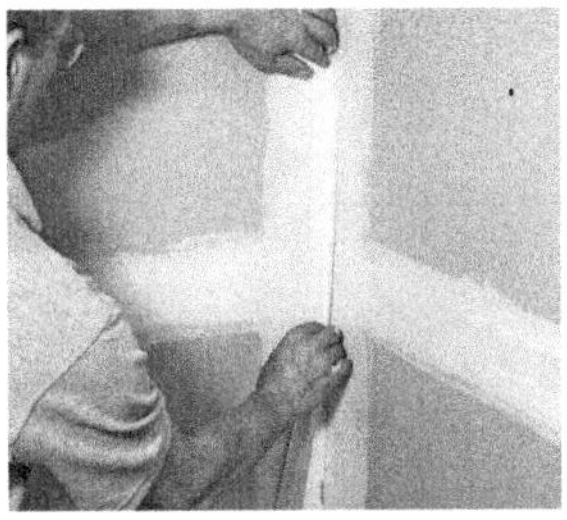

1. ________________________________

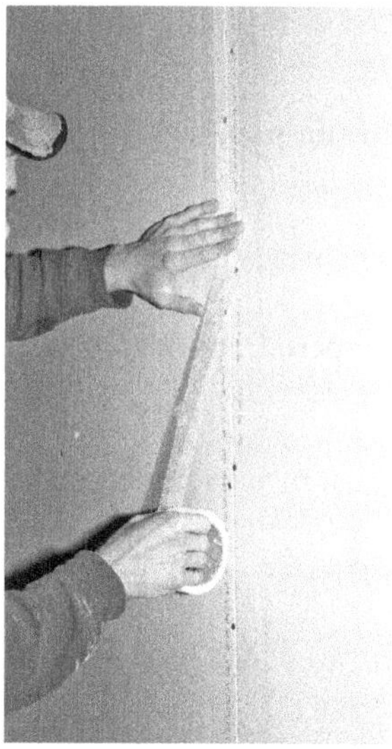

2. ______________________________

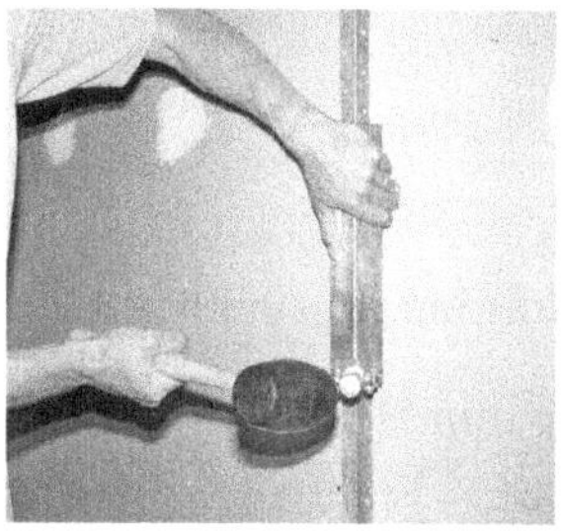

3. ______________________________

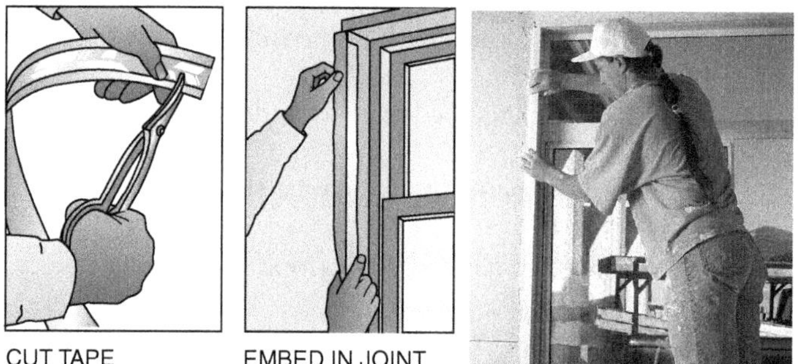

CUT TAPE WITH SNIPS

EMBED IN JOINT COMPOUND

Courtesy of U.S. Gypsum Company.

4. ______________________________

Courtesy of U.S. Gypsum Company.

5. ______________________________

List

List three types of screws used to fasten gypsum panels. Identify the type of material for which the screw would be best used. Write answers on lines provided below.

1. ______________________________

2. ______________________________

3. ______________________________

Procedures

Read the following procedures. Determine what procedure is being described and write the name of the procedure on the blank.

Procedure 1: ____________________

- Plumb the sides of the outlet box down to the floor, or up to the previously installed top panel, by using a framing square or T-square.
- Measure the height from the floor, ceiling, or edge of previous sheet, whichever is more convenient.
- Place the panel in position.
- From the previous layout marks, plumb and measure the outline of the box to be cut.
- With a saw or utility knife, cut the outline of the box.

Procedure 2: ____________________

- Carefully measure and cut the first board to width and length.
- Lift the panel overhead and place in position.
- Hold the board firmly against the framing to avoid gaps.
- Drive fasteners straight into the member.
- Continue applying sheets until the ceiling is covered.

Procedure 3: ____________________

- Prefill joints between panels of ¼ inch or more.
- Center the tape on the joint.
- Lightly press it into the compound.
- Draw the knife along the joint to embed the tape and remove excess compound.
- Apply a thin coat of joint compound.
- Spot coat the fasteners.
- Allow the first coat to dry thoroughly.
- Sand any excess if necessary.
- Feather the second fill coat out about 2 inches.
- Apply a third and finishing coat of compound.

Calculations

Use the information provided to complete your calculations. Write answers on the lines provided.

Estimating wall finish

Estimate the materials for interior wall finish materials for a rectangular 32' × 54' building.

- Drywall sheets are 4' × 12'
- 16" OC framing
- 8' wall height and 140' of interior walls

Estimate the 4 × 8 sheet paneling for one room 16' × 22' with three doors.

Estimate for one room 15' × 18' with 1 × 6 T&G paneling. (See Figure 17-37 in your textbook for area factor.)

Estimate the materials for interior wall finish for a rectangular 28' × 64' building. Drywall sheets are 4 × 12, 16" OC framing, 8' wall height, and 150' of interior walls. One room 14' × 18' with three doors and 4' × 8' sheet paneling. One room 18' × 22' with 10' × 40' T&G paneling (see Fig. 17-37 for area factor).

Item	Formula	Waste factor	Example
Ceiling drywall	building LEN × building WID × waste factor ÷ area per sheet = sheets	5%	64 × 28 × 105% ÷ 48 = 39.2 ⇒ 40 sheets
Exterior Wall Drywall	PERM × wall HGT × waste factor ÷ area per sheet = sheets	5%	184 × 8' × 105%÷ 48 = 32.2 ⇒ 33 sheets
Interior Wall Drywall	wall LEN × wall HGT × 2 sides × waste factor ÷ area per sheet = sheets	5%	150 × 8'× 2 × 105% ÷ 48 = 52.5 ⇒ 53 sheets
Fasteners	NUM sheets × sheet area × 1.5 fastener/SF ÷ fasteners per box = boxes		(40 + 33 + 53) × 48 × 1.5 ÷ 5000 = 1.8 ⇒ 2 boxes
Joint Compound	NUM sheets × sheet area ÷ 1000 × 1.5 pails per 1000 SF = NUM five gallon pails		126 × 48 ÷ 1000 × 1.5 = 9.1 ⇒ 10 pails
Joint Tape	1 – 250' roll per 5 gallon pail		10 pails = 10 rolls
Sheet Paneling	PERM / sheet WID - 2/3 sheet per door = # sheets		2 × 14 + 2 × 18 = 64 / 4 – 2⅔ = 14.6 = 15 sheets
Board Paneling	PERM × room HGT × area factor × 5% waste = board ft	5%	2 × 18 + 2 × 22 = 80 × 8 = 640 × 1.28 × 105% = 860.1 = 861 board feet

Perimeter ______________________

Ceiling drywall ______________________

Exterior wall drywall ______________________

Interior wall drywall ______________________

Fasteners ______________________

Joint compound ______________________

Joint tape ______________________

Sheet paneling ______________________

Board paneling ______________________

CHAPTER 18

Interior Finish

Matching

Match terms to their definitions. Write the corresponding letters on the blanks. Not all terms will be used.

_______ 1. an angled cut starting from the end going back into the face of the molding

_______ 2. the bottom horizontal member of interior window trim that serves as the finished window sill

_______ 3. a piece of the window trim used under the stool

_______ 4. a bevel cut across the width and also through the thickness of a piece

_______ 5. a thin, flat strip of wood inserted into the grooved edges of adjoining pieces

a. apron

b. back miter

c. compound miter

d. hanger

e. main runner

f. spline

g. stool

h. wall angle

True/False

Write *True* or *False* before the statement.

_____________ 1. In remodeling work, a suspended system can be easily installed beneath the existing ceiling.

_____________ 2. Main runners and cross tees should be located in such a way that border panels on both sides of the room are equal and as large as possible.

_____________ 3. Suspended ceiling will have square edges and ends.

_____________ 4. End joints between lengths of ceiling molding may be made square or at an angle.

_____________ 5. Joints on exterior corners are usually coped.

_____________ 6. A miter box is a tool that cuts material at an angle.

_____________ 7. No lines need to be snapped for base moldings or for small-size moldings applied at the intersection of walls and ceiling.

_____________ 8. Door casings are set back from the inside face of the door frame a distance of ¾ inch.

_____________ 9. Prefinished strip flooring is milled with square, sharp corners at the intersection of the face and edges.

_____________ 10. A nail set should be used to set hardened flooring nails.

Multiple Choice

Choose the best answer. Write the corresponding letter on the blank.

_____ 1. Wall angles are _____ pieces that are fastened to the wall to support the other components of a ceiling system.

a. T-shaped c. X-shaped
b. I-shaped d. L-shaped

_____ 2. Main runners are shaped in the form of _____.

a. an upside-down T c. an upside-down L
b. a right-side-up T d. a right-side-up L

_____ 3. Cross tees come in _____ lengths.

a. 1-foot c. 12- and 18-inch
b. 2- and 4-foot d. 36-inch

_____ 4. Slots punched in the side of a main runner are at _____ intervals to receive cross tees.

a. 18- or 24-inch c. 6- or 12-inch
b. 18-inch d. 4-inch

_____ 5. To locate main runners, change the measurement of a room to inches and divide by _____.

a. 48 c. 24
b. 36 d. 12

_____ 6. A suspended ceiling must be installed with _____ inches for clearance below the lowest air duct, pipe, or beam.

a. no less than 6 c. between 2 and 3
b. a minimum of 8 d. at least 3

_____ 7. The beginning ends of the main runners must be cut so that a _____ slot in the web of the runner lines up with the first row of cross tees.

a. suspension c. cross tee
b. main runner d. border

_____ 8. _____ are used to trim around windows, doors, and other openings.

a. Rounds c. Caps
b. Casings d. Backs

_____ 9. _____ molding requires a compound miter.

a. Crown c. Shoe
b. Base d. Chair rail

_____ 10. _____ are small decorative blocks used with molded casings.

a. Plinth blocks c. Back bands
b. S4S stocks d. Stops

_____ 11. Plank flooring comes in various width combinations ranging from _____ inches.

a. 1 to 4
b. 2 to 10
c. 3 to 12
d. 3 to 8

_____ 12. Which of the following oak flooring grades is the best?

a. No. 1 common
b. No. 2 common
c. Select
d. Clear

_____ 13. Which of the following maple flooring grades is the best?

a. Special grade
b. First grade
c. Second grade
d. Third grade

_____ 14. _____ grades of pecan flooring contain all heartwood.

a. First
b. Third
c. Red
d. White

Put in Order

Put in order the steps for cutting a coped joint. Write the corresponding numbers on the blanks.

_____ A. The coped molding end is first cut as if it were an inside miter. The face side forms the profile.

_____ B. Carefully cut along the outlined profile with a fine-tooth coping saw blade, keeping the saw blade plumb.

_____ C. Hold the molding so it is over the end of the sawhorse, and the side of the molding that will fit to the wall is flat on the top of the sawhorse.

_____ D. Cut a molding piece to fit into the corner with a squared end and install.

_____ E. The second piece is cut with a coped joint. Mark a W to indicate which side fits against the wall.

List

List ten standard types of interior molding patterns available. Write the answers on the lines below.

1. ______________________________

2. ______________________________

3. ______________________________

4. ______________________________

5. ______________________________

6. ______________________________

7. ______________________________

8. ______________________________

9. ______________________________

10. ______________________________

Procedures

Read the following procedures. Determine what procedure is being described and write the name of the procedure on the blank.

Procedure 1: ______________________

- Mark the wall using a scrap piece of the material to be installed.
- Apply the pieces to the first wall with square ends in both corners.
- If more than one piece is required to go from corner to corner, the butt joints may be squared or mitered.
- Mitering with the same side down each time helps make fitting more accurate, faster, and easier.

Procedure 2: ______________________

- Hold a piece of side casing in position at the bottom of the window and draw a light line on the wall along the outside edge of the casing stock
- Mark a distance outward from these lines equal to the thickness of the window casing.
- Cut a piece of stool stock to length equal to the distance between the outermost marks.
- Position the stool with its outside edge against the wall.
- Mark the cutout so that, on both sides, an amount equal to twice the casing thickness will be left on the stool.
- Cut to the lines and smooth the sawed edge that will be nearest to the sash.
- Shape and smooth the ends the same as the inside edge.
- Fasten the stool in position using calk and finish nails.

Procedure 3: ______________________________

- Hold a scrap piece on the piece.
- Draw its profile flush with the end.
- Cut to the line with a coping saw.
- Sand the cut end smooth.
- Return the other end upon itself in the same manner.
- Place the piece in position with its upper edge against the bottom of the stool.
- Fasten the piece along its bottom edge into the wall.
- Drive nails through the stool into the top edge of the piece.

Procedure 4: ______________________________

- Check the subfloor for any loose areas and add nails where appropriate.
- Sweep and vacuum the subfloor.
- Determine the direction of the flooring.
- Cover the subfloor with building paper.
- Snap chalk lines on the paper showing the centerline of the floor joists.

Calculations

Use the information provided to complete your calculations. Write answers on the lines below.

Estimating interior finish materials

Estimate the materials for interior finish in the rooms given:

- A 24' × 38' room is to have 2' × 4' grid suspended ceiling with no lights.
- A second 575 SF room is being covered with a wood strip flooring that comes in 15 SF cartons.
- A third 16' × 20' room is being trimmed for windows, doors, base board, and ceiling molding. There is one 4030 window and two 2'-6" × 6'-8" doors.

Estimate the materials for interior finish in the rooms given. A 24' × 38' room is to have 2 × 4' grid suspended ceiling. A second 475 SF room is to be covered with a wood strip flooring that comes in 15 SF cartons. A third 18' × 26' room is to be trimmed for windows, doors, base board, and ceiling molding. There is a 5'-0" × 4'-0" window and two 2'-6" × 6'-8" doors (both sides).

Item	Formula	Waste Factor	Example
Suspended Ceiling Wall Angle	PERM ÷ wall angle LEN = pieces		2 × 24 + 2 × 33 = 114 ÷ 10 = 11.4 + 12 pieces
Suspended Ceiling Main Runner	[room WID ÷ main tee spacing − 1] × [room LEN ÷ main tee LEN rounded up to nearest ½ piece] = pieces		[24 ÷ 4 − 1] × [38 ÷ 12] = 5 × 3.5 = 17.5 ⇒ 18 pieces
Suspended Ceiling 4-foot Cross Tees	[main runner rows + 1] × [room LEN ÷ 2 − 1] = NUM of 4 ft cross tees		[5 +1] × [38 ÷ 2 − 1] = 6 × 18 = 108 cross tees
Suspended Ceiling Hanger Lags	NUM main tees × 3 = pieces		18 × 3 = 54 pieces
Suspended Ceiling Wire (50' roll)	NUM lags × [suspended distance + 2'] ÷ 50 = rolls		54 × [2 +2] ÷ 50 = 4.3 ⇒ 5 rolls
Suspended Ceiling Panels	[NUM 4-foot cross tees + NUM main tee rows + 1 − one per light] ÷ NUM per box = boxes		[108 + 5 + 1 − 12] ÷ 10 = 10.2 + 11 boxes
Window Casing	2 × (width + 1') + 2 × (length 6")		2 (5 + 1) + 2 (4 + 0.5) = 21'
Window Stool	width plus one foot		5 + 1 = 6'
Door Casing	2 × [width + 6" + 2 × (LEN + 4")]		2 × [2.5 + 0.5 + 2 × (6'-8" + 4")] = 34' × 2 = 68'
Ceiling Molding	PERIM		2 × 18 + 2 × 26 = 88'
Base Molding	PERIM minus door widths		88' − (2 × 2.5) = 82'
Wood Strip Flooring	area to be covered ÷ SF per carton × waste = cartons	2%	475 ÷ 15 × 102% = 32.3 = 33 cartons

Perimeter ______________________

Suspended ceiling ______________________

Wall angle ______________________

Suspended ceiling main runner ______________________

Suspended ceiling 4 feet cross tees ______________________

Suspended ceiling hanger lags ______________________

Suspended ceiling wire (50' roll) ______________________

Suspended ceiling panels ______________________

Window casing ______________________

Window stool ______________________

Door casing ______________________

Ceiling molding ______________________

Base molding ______________________

Wood strip flooring ______________________

CHAPTER 19

Stair Construction

Matching

Match terms to their definitions. Write the corresponding letters on the blanks. Not all terms will be used.

_______ 1. the entire stair rail assembly

_______ 2. railing on a stairway intended to be grasped by the hand to serve as a support and guard

_______ 3. upright post supporting the handrail in a flight of stairs

_______ 4. vertical member of a stair rail, usually decorative and spaced closely together

_______ 5. opening in the floor where stairs are located

_______ 6. hole drilled for the thicker portion of a wood screw

_______ 7. sloping portion of trim, such as on gable ends of a building or staircase

a. baluster

b. balustrade

c. handrail

d. newel post

e. nosing

f. rake

g. shank hole

h. stairwell

i. stringer

True/False

Write *True* or *False* before the statement.

____________ 1. Finish stairs extend from habitable level of a house to another.

____________ 2. Open stairways are more economical to build than closed stairways.

____________ 3. The tread run includes the nosing.

____________ 4. The unit run (tread run) is measured from the face of one riser to the next riser.

____________ 5. Decreasing the unit rise decreases the run of the stairs.

____________ 6. A longer stairwell will provide more headroom.

____________ 7. Stair carriages are notched under the risers and treads.

____________ 8. Fasten a temporary riser board about halfway up the flight to straighten and maintain the carriage spacing.

____________ 9. The tread molding is installed under the overhang of the tread and against the riser.

____________ 10. Riser length on an open stair carriage is critical and must be measured precisely.

Multiple Choice

Choose the best answer. Write the corresponding letter on the blank.

_____ 1. A _____ stairway has intermediate landings between floors.

a. platform
b. straight
c. winding
d. stringer

_____ 2. The preferred angle for a staircase is _____ degrees.

a. no more than 25
b. a minimum of 43
c. between 30 and 38
d. approximately 45

_____ 3. The IBC specifies that the height of a riser shall _____ inches and that the depth of a tread _____ inches.

a. not be more than 6; no more than 9
b. not exceed 7 ¾; not be less than 10
c. be between 6 and 8; of at least 8
d. be at least 8 ¼; of less than 10

_____ 4. A _____ are used to dado stringers.

a. coping saw and miter box
b. router and stair jig
c. hammer-drill and laminate trimmer
d. circular saw and saber saw

_____ 5. The nosed edge of the tread projects beyond the face of the rise by _____ inches.

a. 1 ⅛
b. 1 ½
c. 1 ¾
d. 1 ⅝

_____ 6. Fasten tread molding in place with _____ finish nails spaced about _____ inches apart.

a. 3d; 6
b. 3d; 12
c. 4d; 6
d. 4d; 12

_____ 7. A _____ is used to lay out the plumb lines on the open finish stringer.

a. deacon
b. vicar
c. preacher
d. rector

_____ 8. When cutting dadoes using a stair template and a router, the router is equipped with a _____ bit.

a. chamfer
b. cove
c. round over
d. straight

_____ 9. What a balustrade ends against a wall, a(n) _____ newel can be fastened against the wall.

a. second floor
b. half
c. intermediate landing
d. starting

_____ 10. Handrails are typically placed _____ inches vertically above the nosing edge of the tread.

a. 34 to 38
b. no less than 35
c. no less than 37
d. 25 to 45

_____ 11. If square top balusters are used on a balcony, they are inserted into a _____ handrail.

a. plowed
b. mowed
c. squared
d. arched

Terms

Read the following definitions for stair framing terms. Write the name of the correct terms on the corresponding blanks.

_______________ 1. Vertical distance between finish floors

_______________ 2. Total horizontal distance that the stairway covers

_______________ 3. Vertical distance from one step to another

_______________ 4. Finish material that covers the rise

_______________ 5. Horizontal distance between the faces of the risers

_______________ 6. Horizontal member on which feet are placed when climbing or descending the stairs

_______________ 7. The part of the tread that extends beyond the face of the riser

_______________ 8. Main support under the risers and treads

_______________ 9. The smallest vertical distance between the stairs and the upper construction over the foot of the stairs

_______________ 10. Type of stairs in which the treads and risers end against a vertical surface

_______________ 11. Type of stairs in which the ends of the tread and risers are visible

Put in Order

Put in order the steps to finish a closed stair carriage. Write the corresponding numbers on the blanks.

_______ A. Start at the top and work down.

_______ B. Remove the temporary treads installed previously as work progresses downward.

_______ C. Rip the riser stock to the proper width.

_______ D. With the top edge flush with the level cut of the carriage, fasten the risers in position.

_______ E. Cut the risers to length, with square ends, about ¼ inch less than the distance between walls.

List

List the equations needed to determine the rise and run of a set of stairs.

1. ______________________________ = total rise

2. ______________________________ = number of risers

3. ______________________________ = number of treads

4. ______________________________ = total run

Procedures

Read the following procedures. Determine what procedure is being described and write the name of the procedure on the blank.

Procedure 1: ______________________

- Install the finish stringer.
- Cut risers to length, making a square cut on the end that goes against the wall.
- Make miter cuts on the other end to fit the mitered plumb cuts of the open finish stringer.
- Fasten the miters with glue and nails to each stair carriage.
- Rip treads to width.
- Cut one end of the tread to fit against the closed finish stringer and cut the other end to receive the return nosing.
- Apply return nosings to the open ends of the treads.

Procedure 2: ______________________

- On the face side of the stringer stock, lightly draw a line parallel to and about 2 inches down from the top edge.
- Lay out the risers and treads for two or three steps of the staircase.
- Lightly square lines to the top edge of the stringer at the intersection of the face sides of tread and riser.
- With a tape stretched on the parallel line, mark off the unit length for each remaining step.
- Set the template over the first step and adjust it to match the layout.
- Clamp the template to the stringer
- Rout the stringer.
- Loosen, adjust, and reclamp the template to rout the second step.
- Repeat for remaining steps.
- Cut and fit the bottom end of the stringer to the floor and the top end to the landing.
- Make end cuts that will join with the baseboard properly.

Procedure 3: ______________________________

- Cut and rip the required number of risers to a rough length and width.
- Determine the face side of each piece.
- Rip and cut the treads to width and length.
- On the open side of the staircase, where the riser and open stringer meet, make a miter joint so no end grain is exposed.
- Install the risers with wedges, glue, and screws between housed stringers.
- Install treads with wedges, glue, and screws on the closed side.
- Install glue blocks at intervals on the underside of the tread against the backside of the riser to reinforce the corners.

Calculations

Use the information provided to complete your calculations. Write answers on the lines below.

Determine the rise and run of a set of stairs

Determine the rise and run for a set of stairs connecting two floors.

- The distance between the floors is 9 feet
- Unit run is 10 ¼"
- The second floor is framed by using 2 × 10s and has a $^{23}/_{32}$-inch subfloor with a $^{5}/_{8}$-inch wood floor

1. Total rise ______________________________
2. Number of risers ______________________________
3. Number of treads ______________________________
4. Total run ______________________________

CHAPTER 20

Cabinets and Countertops

Matching

Match terms to their definitions. Write the corresponding letters on the blanks. Not all terms will be used.

______ 1. framework of narrow pieces on the front of a cabinet making the door and drawer openings

______ 2. used to apply pressure over the surface of contact cement bonded plastic laminates

______ 3. point of rotation

______ 4. cutout made in a piece to receive another piece, such as a cutout for a butt hinge

______ 5. guide on the end of edge-forming router bits used to control the amount of cut

______ 6. method used to bend plastic laminate to small radii

a. base unit

b. face frame

c. gain

d. J-roller

e. pilot

f. pivot

g. postforming

h. toe space

True/False

Write *True* or *False* before the statement.

__________ 1. European-style cabinets do not have face frames.

__________ 2. Most base cabinets are manufactured 36 inches deep.

__________ 3. Countertops should be made of ½ inch solid wood.

__________ 4. Vertical-type laminate is relatively thin—about ⅛ inch.

__________ 5. Using a table saw is the easiest method for cutting laminate.

__________ 6. Contact cement is used for bonding plastic laminates.

__________ 7. Strips of laminate can be cold bent to a minimum radius of about 6 inches.

__________ 8. The semi-concealed offset hinge is designed so that only the pin is exposed when the door is closed.

__________ 9. Drawer sides and backs are generally ½-inch thick.

__________ 10. Dovetail joints are used in higher-quality drawer construction.

Multiple Choice

Choose the best answer. Write the corresponding letter on the blank.

_____ 1. The surface of a countertop is usually _____ from the floor.
a. about 36 inches
b. 35 inches or less
c. 34 to 38 inches
d. approximately 32 inches

_____ 2. Wall units are installed _____ inches above the countertop.
a. approximately 20
b. about 18
c. no less than 24
d. about 30

_____ 3. The _____ cabinets are for use in kitchens without soffits.
a. 15-inch
b. 18-inch
c. 24-inch
d. 42-inch

_____ 4. The recess at the bottom of a cabinet is called a _____ space.
a. shoe
b. standing
c. toe
d. foot

_____ 5. _____ pieces fill small gaps in width between wall and base units when no combination of sizes can fill the existing space.
a. Joiner
b. Filler
c. Link
d. Connector

_____ 6. Postforming is bending the laminate with heat to a radius of _____.
a. ¾ inch or less
b. more than ¾ inch
c. more than ½ inch
d. ⅝ inch or more

_____ 7. If a backsplash is used, rip a _____ -inch wide length of _____ -inch stock the same length as the countertop.
a. 6; ⅝
b. 4; ¼
c. 5; ½
d. 4; ¾

_____ 8. Heating laminate to _____ degrees Fahrenheit uniformly over the entire bend will facilitate bending to a minimum radius of about 2 ½ inches.
a. 150
b. 220
c. 325
d. 475

_____ 9. The _____ type of door laps the entire thickness of the door over the opening, usually ⅜ inch on all sides.
a. overlay
b. lipped
c. flush
d. lapping

_____ 10. _____ hinges have leaves that are set into gains in the edges of the frame and door.
a. Pivot
b. Butt
c. Case
d. Surface

Terms

Read the descriptions of types of cabinets. Write the cabinet type on the corresponding blanks.

_______________ 1. Standard cabinets are 12 inches deep; standard height is 30 inches; typically range in width from 9 to 48 inches in 3-inch increments

_______________ 2. Manufactured at 34 ½ inches high and 24 inches deep; manufactured widths range from 9 to 24 inches (single door) and 27-48 (double door); bottom is recessed

_______________ 3. Usually manufactured 24 inches deep; made 66 inches high and in widths of 27, 30, and 33 inches

_______________ 4. Made 31 ½ and 34 ½ inches high and 16 to 21 inches deep; widths range from 24 to 36 inches in increments of 3 inches then 42, 48, and 60 inches

Put in Order

Put in order the steps for laying out cabinet lines. Write the corresponding numbers on the blanks.

_______ A. At each stud location, draw plumb lines on the wall.

_______ B. Measure 34 ½ inches up the wall and draw a level line to indicate the tops of the base cabinets. Measure and mark another level line at 54 inches from the floor, to where the bottom of the wall units are installed.

_______ C. Mark the outlines of all cabinets on the wall to visualize and check the cabinet locations against the layout.

_______ D. Above the upper line on the wall, drive a small nail in at the point where the wall is solid to accurately locate the stud.

_______ E. Mark the locations of the remaining studs where cabinets will be attached.

Identify

Identify the object or the activity depicted in the images below. Write your answer on the line next to the image.

1. ________________________________

2. ________________________________

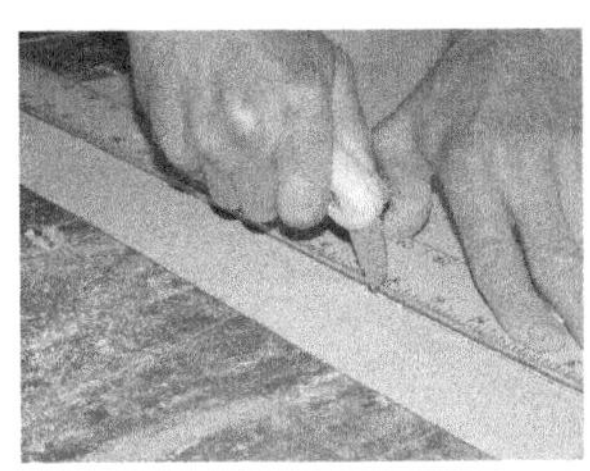

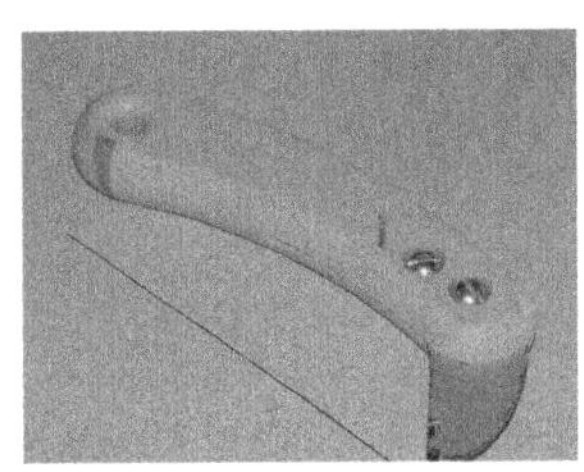

3. ______________________________

4. ______________________________

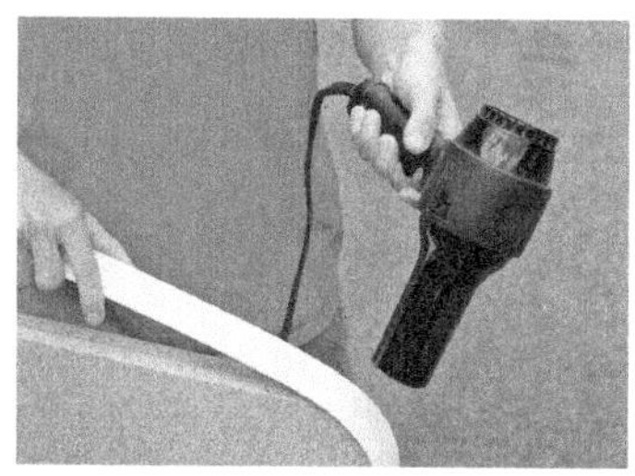

5. ______________________________

6. ______________________________

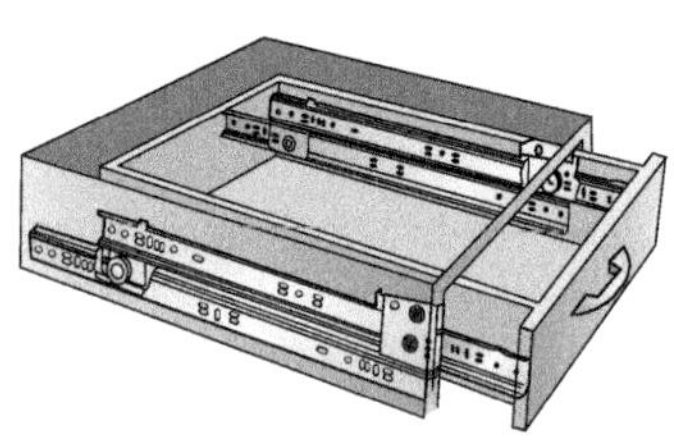

Courtesy of Knape and Vogt Manufacturing Company.

7. ______________________________

List

List three typical joints between a drawer side and back. Write the answers on the lines below.

1. ______________________________
2. ______________________________
3. ______________________________